Microcomputer Algorithms

Action from Algebra

Microcomputer Algorithms

Action from Algebra

J P Killingbeck

Adam Hilger
Bristol, Philadelphia and New York

British Library Cataloguing in Publication Data

Killingbeck, J. P. (John Patrick) 1938–
Microcomputer algorithms.
1. Algorithms. Applications of microcomputer systems
I. Title
511.80285416

ISBN 0-7503-0097-3

Library of Congress Cataloging-in-Publication Data

Killingbeck, J. P. (John P.)
Microcomputer algorithms: action from algebra/J. P. Killingbeck.
250p. 24cm.
Includes bibliographical references and index.
ISBN 0-7503-0097-3 (pbk.)
1. Algebra–Data processing. 2. Microcomputers–Programming.
I. Title.
QA155.7.E4K53 1991
512′.00285–dc20 90-22739

Published under the Adam Hilger imprint by IOP Publishing Ltd
Techno House, Redcliffe Way, Bristol BS1 6NX, England
335 East 45th Street, New York, NY 10017-3483, USA
US Editorial Office: 1411 Walnut Street, Suite 200, Philadelphia, PA 19102

Typeset by KEYTEC, Bridport, Dorset
Printed in Great Britain by J W Arrowsmith Ltd, Bristol

Contents

Preface

Perhaps the first thing I should do is to explain the subtitle of this book, since it is intended to represent the principal point of view from which the book is written. The subtitle is actually a punning one. It carries the obvious exhortation: let's get this abstract mathematics to do something useful! However, the word *action* is used by writers on systematic computer programming (e.g. Wirth 1973) as a technical term, in the sense that an algorithm specifies some sequence of actions to be performed on appropriately defined data objects in order to transform some set of inputs into a set of outputs. Within that context, the subtitle also refers to this book's central theme, that in original scientific and mathematical computations it is the preliminary derivation of the relevant algebraic formulae and their representation by a sequence of actions within some algorithm which takes up most of the effort. Producing the final realization of the algorithm in some *particular* programming language is the last step in the process. It is, of course, a vital step; nevertheless, I incline to that view of the computing process which is sometimes crudely expressed by saying that 90 per cent of 'programming' is done without actually touching the keyboard.

Having read scores of books and papers by the proponents of particular high-level languages and of structured programming in general, I have concluded that the greatest wisdom is to be found in the works of those authors (e.g. Guttmann 1977, Gries 1981, Grogono and Nelson 1982, Ghezzi and Jazayeri 1982, Harel 1987) who either stress the language-independent features of algorithm construction or survey objectively the typical actions on data which a useful high-level language should be capable of representing. I have been pleased to observe that this reasonable mathematical and logical approach is taken, in their more sober moods, even by those writers who are notorious for denouncing the thought-corrupting evils of any high-level language which they did not personally help to devise. It is a truism that the way in which mathematical equations are manipulated or considered can be conditioned by the knowledge that some computer language is to be used to implement any resulting algorithm. However, the constraints imposed by this knowledge usually lead the programmer to think in terms of suitable modular, subroutine and loop structures to realize the required sequence of actions. Such concepts are really structural (rather than language-specific) ones, and any of the common high-level languages can implement them with ease. There are, of course, important distinctions between parallel and sequential algorithms (e.g. Gibbons and Rytter 1988) and between algorithms for imperative and functional languages (Ghezzi and Jazayeri 1982). However, these distinctions are not of principal concern within this book, which uses the simple sequential imperative language BASIC for the realization of the algorithms produced by the mathematical theory.

I can best explain how my attitude towards the use of computers in mathematics and science has evolved by means of two anecdotes.

A few years ago, I attended a gathering of 15 or so physicists, mathematicians and computer scientists who had been invited to discuss the future provision of computer programming courses within a certain university. The newly arrived professor of computer science immediately announced that PASCAL was really the only respectable language and that FORTRAN courses should be phased out. A shocked silence followed, until I quietly pointed out that many scientists still found FORTRAN useful because of its well developed NAG library and other features, including complex number operations, which standard PASCAL could not match at that time. After that, the other scientists revived, and eventually obtained an agreement that FORTRAN courses would continue. My neighbour at the table remained silent throughout the debate, but then whispered 'I daren't say anything; I use BASIC'. 'So do I', I replied, 'except when I'm using a programmable calculator. I just wanted to make the FORTRAN boys stick up for themselves'. If that professor were to arrive today he would perhaps tell us that MODULA 2 or ADA is the in-thing which must displace everything else—and I would, if present, lead the same commonsense rebellion which I led then.

My second story concerns an invited talk which I gave on microcomputer algorithms, at a northern university. I demonstrated two scientific computations which gave results of such high accuracy that they compared favourably with previously published results obtained using mainframe computers. One professor asked: 'Do you do all this original work using only a BBC microcomputer?'. 'Only on a bad day', I replied, producing laughter from the audience: 'I usually use a Sinclair Spectrum' (then Britain's most popular home computer). After the talk I was mildly chided for undermining the case, then being made by my hosts, for the purchase of a large computer. 'You give big computers a bad name' I was told. 'No', I replied, 'I just give thinking a good name'. The point is that there are many tasks which *need* the speed and capacity of a mainframe computer, not least the task of providing an interactive computing environment for an entire class of students simultaneously. Further, any algorithm which works efficiently on a microcomputer will almost certainly transfer advantageously to a mainframe computer; careful thinking will only serve to enhance the effective power of a computer of *any* size. My hosts thus had every reason to continue their pressure for a new and larger computer, since I was merely illustrating the value of the microcomputer as a stimulus to that individual creativity which ultimately benefits all computer users.

The moral of my first story is: if you are primarily concerned with the actual *use* of computers in creative mathematical or scientific work, don't be too impressed by experts in computer science. The best way to debunk some of their more dogmatic assertions is by example rather than by polemics. I still find it unfailingly risible to hear or read that BASIC is 'not suitable for serious programming', or that it 'lacks structure'. It has always been my experience that in mathematical work the mathematics itself usually produces sufficient structural backbone to be going on with, and that BASIC is sufficiently simple and flexible to conform to any required pattern imposed by the mathematics. The evidence of the Science Citation Index suggests that a sizeable body of opinion regards my own work using BASIC as reasonably serious and original, and the more extreme criticisms of BASIC are really rather insulting to the serious educators who devised it. However, the best contribution which I can make to the debate is to state that this entire book is my argument.

The reader will by now have gathered that I am unrepentently 'old fashioned'. In my own computing career I started with (and still help to teach) FORTRAN, and have also performed original scientific work on programmable calculators, which certainly force the user to keep a tight check on the assignment of variables to the storage locations. I have used several high-level languages on microcomputers. My personal experience is that the widespread availability of inexpensive microcomputers with BASIC interpreters has given a significant stimulus to my mathematical and scientific creativity, and I believe that I am not alone in that experience. Given the millions of domestic microcomputers in existence, I can only regret that the tremendous potential which they represent for scientific and mathematical education does not yet seem to have been fully realized.

It is fast becoming the practice for books which set out microcomputer programs to have accompanying software packages, and I had to consider whether to follow this trend. My current view is that if programs are short (as mine are) then the most *universal* form of presentation is a printed listing, which can later be converted into a stored program for any particular microcomputer. The action of typing in each program then becomes part of the process of understanding the program's structure and the way in which it is related to the underlying mathematical theory. I am supposing that my readers will be sufficiently serious to expend a little careful effort in the cause of learning, and also that they will understand the notion of fair play; the work is formally protected by copyright, although my intention is that plentiful and productive use should be made of the programs by those who purchase the book. To use the programs without first reading and understanding the book's account of how they arose and were constructed is to miss out the experience of invention and discovery which I have tried to share with the reader. The critical and mathematical study of algorithms and programs is something which I have always been keen to encourage in my own students in this era of 'black box' programs. I regard it as important for the scientific and mathematical academic community to remind students continually of the need to have a critical understanding of the tools which they use (or invent); this is particularly necessary at a time when the British university scene seems to be in danger of domination by some of the more philistine elements in applied science and engineering.

In the preface to my book on microcomputer quantum mechanics (Killingbeck 1985a) I commented that the topics which I covered were in several respects complementary to those covered by John Nash in his book (Nash 1979) on computational methods for small computers. That comment still holds with respect to this book and the second (1990) edition of his book; the two Hilger books between them cover a wide range of ideas and methods of value to scientists and mathematicians who wish to use microcomputers in research and education. Wherever I have treated the same topics as Nash (e.g. the generalized inverse) I have tried to use a different point of view or an alternative algorithm, to illustrate that even in numerical work there is some latitude for personal style and taste in the choice of methods. This book shows some bias towards techniques which are useful in quantum mechanics; this is particularly clear in the last chapter, which sets out two of my research programs for the Schrödinger equation. However, I estimate that some two-thirds of the programs in this book are directly useful in other areas of applied mathematics. Even those which deal explicitly with quantum mechanics are explained in such a way as to give the reader useful background information about methods (e.g. the finite difference method) which are valuable as computational tools in many other areas.

It is often taken to be a formal but obligatory ritual that an author's spouse should be thanked for patience and forbearance. In my case such thanks are no mere formality, for my wife has had the patience to produce the manuscript for me as fast as I could write the material. I would like to thank Maureen Clarke of Adam Hilger for her tolerance in waiting for the revised chapters which I was somewhat tardy in writing, and also Neville Goodman, who years ago recruited me as an author for Adam Hilger. My final thanks go to someone who probably does not expect them and whom I have never met. He was responsible for making the microcomputer popular throughout Britain, and under his auspices I received assistance in my early research and teaching applications of microcomputers. His version of BASIC turned out to be splendidly suited to my own style of scientific computing, and the original programs described in this book, although they are simple enough to work with any system of microcomputer BASIC, were actually put together and tested on a Spectrum 48K machine of 1982 vintage. Thank you, Clive Sinclair, sir.

Low Drewton Cottage
Station Road
South Cave
North Humberside HU15 2AE

John Killingbeck
June 1990

PS Old-fashioned Spectrum folk can write to me to hear about the high-speed compiled versions of my programs, used in my research on the old-style 48K Spectrum.

1 General introduction

1.1 The microcomputer style of programming

Modern microcomputers tend to be faster in operation and to have much larger memory capacities than their predecessors. It is sometimes claimed that this increase in power has made it obsolete to refer to microcomputer methods as being something distinct from computer methods in general. The view adopted throughout this book is one which has been expressed elsewhere (Killingbeck 1985a, b), namely that there *is* a style of producing algorithms and programs which is specially appropriate for microcomputers, and that this style manifests itself both in the form of the algorithms and in the way in which the operator uses the resulting programs based on the algorithms.

The most obvious aspect of a specialized style connected with microcomputer work is the tendency to look for algorithms which are based on iterative methods and the use of recurrence relations. Such algorithms lead to programs which do not occupy much memory space, since they involve loop structures in which a basic set of simple actions is repeated until some desired result has been obtained. The amount of memory space occupied by the data which are manipulated by the program is also often reduced by various stratagems. Very often a recurrence relation algorithm will only use a few elements of an array at a time, and it may be necessary to store only the currently active elements, so that previous elements which do not contribute to the final result can be discarded once they have been used in the recurrence relation to give the values of later elements.

When an algorithm is actually realized in the form of a microcomputer program in BASIC, as are the algorithms of this book, then the interactive capabilities of the language can be used as part of the process of performing the computation. If the program is composed of modules, each of which performs some definite known set of actions on the data, then manual GO TO commands from the keyboard can be used to perform those actions in an order which augments or varies the automatic sequence imposed by the stored program. For example, the value of a variable can be changed by going to an input module, while leaving all the previously computed variables with their current values; to use RUN to reset the variable would usually initialize all variables to zero. In a program such as GENFIT in this book, the operator can voluntarily use a GO TO command to refine a matrix inverse if he is doubtful about its accuracy. A general feature of interactive interpreted BASIC, as available on most microcomputers, is that it allows the operator to participate in the computation; the calculating speed of the electronic circuitry is allied with the experience and judgement of the human operator. The BASIC command which allows this is GO TO, which is sometimes criticized by writers on structured programming. As several authors have made clear (e.g. Knuth 1974, Sale 1975, Yourdon 1975, Killingbeck 1985b) the fundamental concepts in structured programming are those of sequencing, alternation and repetition of processes. These are quite general concepts, applicable in any high-level language. The *particular* control instructions which are advocated in connection with structured programming (e.g. IF–THEN–ELSE, DO WHILE) can be realized in a form which includes GO TO instructions; some of the authors cited above give specific examples. The essential point is that if a program is structured into clearly marked blocks of code which perform specific actions in the manner of black boxes, then the presence or absence of explicit GO TO instructions in the innards of each black box is a matter of secondary importance. This slightly more abstract notion of what 'structure' means is natural to anyone used to mathematical reasoning, and a more formal mathematical approach to the formulation of computer programming principles has been advocated by several writers (Hoare 1969, 1986, 1987, Mills 1975, Zave 1984).

1.2 Modules and subroutines

As implied by its subtitle, this book is based on the idea that an executed program carries out a sequence of actions, each of which corresponds to a set of well defined mathematical operations in the underlying algebra on which the algorithm and program are based. It is thus natural to construct a program from modules, each of which carries out some clear action on the variables and data. This way of constructing a program obviously leads to the notion of a procedure, and it is often claimed that those forms of BASIC (e.g. BBC BASIC in Britain) which permit the use of named procedures represent a marked advance over earlier forms of BASIC. As far as advances in BASIC are concerned, Shammas (1988) pointed out that some recent versions of BASIC compare favourably with PASCAL, MODULA 2 and C in terms of providing commands which express directly the operations required by a structured programming approach. However, in the writing of this book it has been taken as necessary to aim for some degree of universality; from that point of view the most important feature of more advanced forms of BASIC is that they are downwards compatible, in the sense that most of them still retain the simple GOSUB and GO TO commands. Although a subroutine (unlike a procedure) is invoked by transferring control to a specific numbered line in a BASIC program, the use of subroutines in a modular program structure is not difficult to achieve if a few conventions about line numbering are established and adhered to while constructing a suite of BASIC programs.

In this book the following conventions have been followed in most of the displayed programs.

Lines 10 to 200. Input module, for declaration of arrays, input of matrix elements, etc.

Lines 300 to 500. Root-finding module, for use in programs which arise from algorithms based on the root-finding approach.

Line 1000. Subroutine start line for function evaluation subroutines.

Line 2000. Start line for ancillary modules which can be attached to a variety of other programs.

The conventions listed above have been followed in the main, and it will be obvious to the reader when (particularly for very short programs) they have been ignored. An over-riding general principle which has been used throughout is that each group of coherent actions within a program should start with a new hundred as its initial line number and should be marked by a REM line of asterisks, together with a specific name if the actions are sufficiently important to constitute an identifiable module. These conventions in the listed program make it easier to see the structure of the program and also to use GO TO commands if the operator wishes to use the program in an interactive manner. In some programs REM lines which involve commands have been included, as a reminder of alternative actions which could be performed by the program if desired.

By reserving particular line number regions for particular types of action, it becomes possible to use some modules as constituents of several different programs. This particularly applies to the root-finding modules and to the ancillary modules which perform Padé approximant or Richardson extrapolation analyses of the output from the main program. What the modular structure means is that most of the programs of this book do not only carry out some particular computation. They also serve as computational test beds, and can be modified, extended or restructured in various ways once the underlying mathematical theory and algorithm have been understood. By appropriate allocation of actions to modules it is often possible to ease the way for subsequent modifications and extensions of a program. The programs in this book may well be capable of further improvement; they should certainly be capable of providing project material for science students interested in computation.

Several varieties of BASIC provide the facility to load a stored program so as to blend with a program already in the microcomputer. This facility, called MERGE in Sinclair BASIC and SPOOL in BBC BASIC, allows prestored modules to be fed directly into any program in which they form a component, provided that the line number conventions adopted are suitable, since incoming program lines replace existing ones which have the same line numbers. This means that it is often possible to save a sizeable fraction of the effort of constructing a program, since only the entirely new portions of the program need to be typed in manually from the keyboard. The high-level language FORTRAN, as applied in mainframe installations, is often preferred by scientists because of the availability of standard subroutines or procedures (from collections such as the NAG library) which can be incorporated in their programs. From the comments of this section it should be clear that, by following a few simple conventions, users of microcomputer BASIC can build up their own library of reusable subroutines and modules. The September 1984 issue of *IEEE Transactions on Software Engineering* included an informative set of articles dealing with the topic of reusable programs and modules. Myers (1975) gave a valuable analysis of the role of modules in program design.

1.3 Notation and conventions of this book

In the history of programming languages ALGOL played a role in some ways analogous to that of Latin in the history of European languages. It provided notation and concepts which have permeated programming theory in general as well as giving rise to the particular ALGOL–PASCAL–MODULA line of descent (or ascent). Simple BASIC does not have many of the explicit features included in ALGOL, but in the planning of a BASIC program it is still possible to use some of the ideas embodied in the ALGOL structures, as already noted in §1.1. The notion of orthogonality (Tanenbaum 1976) in a language (i.e. the capacity to build up complicated constructions by combining a small number of basic ones) is one which appeals naturally to computer users interested in mathematics and logic. Varieties of the intermediate program description languages (PDL) used in setting out an algorithm prior to coding are similar to ALGOL in their structure; Beech (1980) used a PDL to describe algorithms which he then implemented in BASIC, showing how this two-step approach helps to produce programs which conform reasonably well to modern notions of structured programming.

In describing an algorithm the ALGOL assignment statement symbol := is sometimes used in this book, so as to stand out in contrast to the ordinary equality sign = which has its usual mathematical meaning. Within the textual discussion the distinction between equality (=) and assignment (:= or ←) requires a special symbolism, since the ordinary = sign is also used in the assignment statements of BASIC. Several forms of BASIC allow the use of = without a preceding LET in assignment statements, but most of them are downwards compatible in that they *do* allow the use of LET if the programmer so wishes. The listed programs of this book use LET (which is mandatory in Sinclair BASIC), but they also use Boolean expressions, in which = has its traditional mathematical meaning. For example, the following line represents a legitimate program line in BASIC

```
IF R=1 THEN LET P=P+1                                    (1.1)
```

with the two = symbols having different meanings. An alternative line producing the same results would be

```
LET P=P+(R=1)                                            (1.2)
```

where the Boolean expression in the bracket is given the numerical value 1 (for true) or 0 (for false) by most forms of microcomputer BASIC. That convention is used in the program listings of this book; the reader will be able to make appropriate changes for forms of BASIC which use a different convention. BBC BASIC assigns the numerical value −1 to a true Boolean expression, and *could* have been covered by using $(R=1)*(R=1)$ in the program line (1.2); however, it has been supposed that the reader who understands the purpose of each action in a program will have no difficulty in making the appropriate slight modifications. For quantum mechanical problems the use of Boolean expressions is particularly useful in the computational treatment of the Kronecker delta symbol and the Dirac delta function, as is illustrated by some of the programs in this book. A BASIC line such as (1.1) obviously avoids any problem about interpretation of Boolean functions, but there is still a need to state a convention for IF–THEN lines such as

```
IF R=1 THEN LET P=P+1:LET R=2                            (1.3)
```

The most common convention (used in this book) is that the two statements which follow THEN, and which are separated by the usual statement separator :, are *both* executed if R equals 1 and both not executed otherwise. Another (less common) convention is that whereby R would acquire the final value 2 whatever its value before the line is executed (although it would have to have *some* assigned value to avoid an error signal and a program halt). One way to avoid the problem is to exploit the presence of line numbers in BASIC and use that wicked GO TO command as follows

```
70 IF R<>1 THEN GO TO 80
75 LET P=P+1:LET R=2
80 (rest of program)
```

(1.4)

with starred REM lines at 69 and 76, if desired, to highlight that the two lines together produce a particular action on the values of the variables. Provided that the actions executed by each block of program are clear, there is no great problem about using GO TO commands either within or between them, and the programs of this book use them both in the listed programs and in the suggested mode of use of the programs. The reader is recommended to look at the ACM Forum section of issues 3, 5 and 6 of the 1987 volume of *Communications of the Association for Computing Machinery* to see what is probably the most exhaustive debate ever published about the merits and defects of the GO TO command.

The common BASIC symbol : is used to separate statements which follow one another in a program line. Multi-statement lines have been used in this book to shorten the length of the listings; almost all BASICS permit this. The author has only come across one exception, the BASIC built into the TI99/4A microcomputer, which has the compensating advantage of giving high-precision arithmetic. Because of the intervals of 5 or 10 between consecutive line numbers, it is possible to spread out a multi-statement line to occupy several lines if required.

By devices such as those outlined above it would be possible to write BASIC programs to be universally transportable between all dialects of BASIC. The approach used in this book is to use only very simple operations and commands which are available in most BASICS and to explain the programs in such detail that the reader will find it easy to make slight modifications if he or she uses a particularly restrictive form of BASIC. That this remark has to be made at all serves to illustrate the point that the most universal way to present short programs is via 'listing plus documentation', since it would be difficult to provide software for the full range of microcomputers available to potential readers of this book.

1.4 The program presentation format

Except for the very simple programs, each program in this book is described in a format which includes

1. *The mathematical theory.* This sets out the relevant algebra on which the computation is based, and the way in which it leads to an algorithm.
2. *The programming notes.* These give comments about the problems which arise when the algorithm is to be implemented in the form of a program. Many of these problems are not peculiar to the use of any particular programming language, although they are described in a manner appropriate to the 'microcomputer style' outlined in §1.1.
3. *The detailed program analysis.* This gives a detailed (often line-by-line) description of the listed program, indicating how it embodies the ideas set out in the preceding sections.
4. *The program.* The listing includes asterisked REM lines which box-in various sections of the program, particularly those which form a definite module or subroutine or perform some important action on the variables. Important sections of the program are given a specific title which indicates their role in the computation. A respectable saving in time and memory space can be made by omitting all REM lines, although their inclusion helps to display the structure of the program.
5. *Specimen results.* The results section gives typical results produced by the listed program, so that the reader can check his or her own operation of the program. It also suggests, for some programs, how modifications can be made which use the program modules to perform a different calculation, perhaps involving an algorithm which is different from that on which the original program is based. Ways in which the reader can experiment with or generalize the program are sometimes suggested, and references are given to published applications and to other work relevant to the problem which the program is designed to handle.

1.5 Brief survey of chapter contents

Each chapter has an introductory section which explains any special features which it contains, and so we give here only a brief list of the contents of the later chapters.

Chapter 2 explains the root-finding approach which plays an important role throughout the book. It also sets out some specific root-finding methods and explains how the particular problem of calculating quantum mechanical expectation values can be handled within the context of the root-finding approach.

Chapter 3 sets out the basic theory of Richardson extrapolation, another technique which is used throughout the book. It gives a program which will carry out the extrapolation process and also deals with the process of numerical quadrature, which is the best known example for which Richardson extrapolation is useful.

Chapter 4 deals with several methods of interpolation and extrapolation, including Lagrange interpolation (in Newton's form), quadratic spline interpolation, and the Padé approximant method of extrapolation by rational functions. It also uses quadratic spline functions to illustrate the collocation approach to the Schrödinger equation.

Chapter 5 treats the problem of finding the inverse of a square matrix and the generalized inverse of a rectangular matrix, with particular reference to the least-squares fitting of a polynomial to a set of data points.

Chapter 6 sets out three different methods for the solution of matrix eigenvalue problems, one of them being particularly suitable for the generalized eigenvalue problem and another for the treatment of large matrices in which the matrix elements are given as calculable functions rather than as numerical elements.

Chapter 7 explains two perturbation techniques, both of Rayleigh–Schrödinger type. One of them makes use of the hypervirial theorems to handle a quantum mechanical problem, while the other is more widely applicable to the matrix eigenvalue problems.

Chapter 8 describes two finite difference methods which can be allied with matrix eigenvalue or shooting techniques, together with Richardson extrapolation, to find the eigenvalues of the Schrödinger equation.

Chapter 9 is concerned with methods of eigenvalue calculation which use recurrence relations together with the root-finding approach. Two of the methods involve a power series approach, while the third uses a moment recurrence relation.

Chapter 10 applies a combination of the techniques explained in previous chapters to calculate the energy levels of a perturbed oscillator in two dimensions and of a hydrogen atom in a magnetic field in three dimensions.

2 Root-finding methods and their application

Programs

ROOTSCAN, NEWTON, SECANT, ZIGZAG, SEARCH, MAXMIN.

2.1 General introduction

This chapter sets out that general approach to computational problems which is based on the use of program modules which can find the real roots of scalar equations of the form $F(x) = 0$. How each particular problem is converted into a root-finding problem is explained in detail in those later chapters which use the simple modules described in this chapter. The root-finding methods used are very simple, since the functions produced by the problems of this book are for the most part well behaved. The main features to be noted are that all the methods are iterative (leading to loop program structures) and that the shift SH on each cycle is evaluated, so that it can be subjected to numerical tests before it is added to the current root estimate. Since the roots of the derivative of a function correspond to the stationary points of the function, there is a link between the problems of finding roots and of finding extrema. This point is discussed in §2.6, which gives a somewhat unusual application of Richardson extrapolation. Sections 2.9 and 2.10 point out how the ideas of perturbation theory make it possible to exploit the root-finding approach to calculate quantum mechanical expectation values; this strategem is used at several points throughout the book.

2.2 The root-finding approach

One of the techniques consistently used throughout this book is that of transforming a piece of mathematical theory so that the desired solution to a problem can be obtained by finding the roots of some equation of the type $F(x) = 0$. The function $F(x)$ might not be a simple one. It is often the output of a subroutine for which x is the input, so that F is produced by some complicated algorithm and cannot be written down as a simple explicit formula. Two examples of this are provided by the programs FOLDER and FIDIF; in the former the function $F(x)$ gives the determinant of a matrix, while in the latter it gives the value of the wavefunction at the upper limit of the region of integration.

One of the special advantages of the root-finding approach is that, whenever $F(x)$ has a root, then so does any multiple of $F(x)$. This makes it possible to apply numerical scaling at various points in the computation (in order to avoid overflow or underflow problems) without changing the solution to the problem. This invariance under numerical scaling is a feature which can be obtained with particular ease for problems in the area of quantum mechanics or linear algebra in general. That is why the present author was initially drawn to the root-finding approach, which has been described in other contexts in books such as that by Rice (1983) and that by Kantaris and Howden (1983).

The root-finding approach lends itself particularly well to a modular style of programming. Given a root-finding module which makes as part of its operation a call to a subroutine in order to produce $F(x)$ for a given x, all that is necessary to change the problem is to change the subroutine (i.e. to use a different function). This means that the essential task in each case is one of mathematical analysis; the problem to be solved is first transformed into a form where it becomes a root-finding problem of the type $F(x) = 0$ and then the final step is to write a subroutine which produces the appropriate output $F(x)$ when the root-finding module passes it the value of x. The root-finding module can be modified (e.g. to implement Newton's method or the secant method or the bisection method) if it turns out that the function $F(x)$ has a form which is particularly suited to one of the traditional root-finding techniques.

The functions $F(x)$ produced by the programs of this book fall into two principal types. The first type has single zeros separated by extrema, while the second has a singularity between each pair of zeros (MOMOSC produces such a function). For want of better names we can term these function types the cos type and the tan type, respectively. The root-finding methods which we use are designed to work effectively for these two types of function. The applications of the root-finding approach spread beyond the areas covered in this book, in which we mainly describe cases for which real single roots of scalar functions are required. Even within this restriction several problems of quantum theory and matrix theory can be handled quite easily.

In discussing the operation of root-finding programs and function subroutines, the symbol E rather than x is used in several of the discussions of this book. This obviously makes no difference to the mathematics, but arose because E is often the symbol actually used in programs for which the roots of a function correspond to the energy levels of a system. E is then the trial energy which has to be varied to produce the result $F(E) = 0$.

2.3 Root-finding modules

The use of the root-finding approach to the solution of problems of various kinds means that simple routines for locating the roots of a scalar function will be incorporated in many different programs. In this section we describe some simple routines which can be used as components in other programs in this book. Since the routines are short and easy to follow, they can be described briefly without needing the detailed section-by-section treatment used for the longer programs in the book.

ROOTSCAN

This routine presupposes that a starting value E0 and a starting increment DE have been specified (usually by operator input). It then calls the function subroutine (taken to be at line number 1000) whenever a function value F(E) is required. The contents of the function subroutine are what determine the particular calculation which the overall program is performing. ROOTSCAN starts at E = E0 and then increases E in steps of amount DE, testing to find whether the function F has different signs at E and E + DE. The values of E and E + DE are called E1 and E2 and are stored (along with the function values), since they can be used later in two ways. First, when the routine passes control to the detailed root finder it is helpful to give that routine a good start by using the data stored by ROOTSCAN to estimate the root position by linear interpolation. Second, when the root has been calculated and displayed, E can be set equal to the stored E2 value in order to resume the scan at the point where it was halted.

ROOTSCAN proceeds by looking for changes of sign in F(E) and thus is appropriate for finding simple roots which have a reasonable separation (larger than DE). If F has tan-type singularities, these will also give a change of sign, but will lead to F values which increase as the 'root' is approached. The stored E1 and E2 and the interpolated E value can be used to reject singularities if desired. F(E) will be smaller in magnitude than F(E1) and F(E2) for a normal root, but it will exceed at least one of them for a tan-type singularity. An extra line can be inserted in a program to test for this and continue the scan if the sign change is due to a singularity. The crucial decision for the operator is the choice of E0 and DE, but DE can be decreased if it seems likely that two sign changes are taking place within the interval DE. For the kind of quantum mechanical bound state problems used as examples in this book this kind of difficulty is easily avoided.

NEWTON

The traditional iterative formula of Newton's second-order process for finding roots involves the derivative of the function:

$$E' = E - F(E)/F^{(1)}(E) \tag{2.1}$$

The *explicit* statement of both functions F and $F^{(1)}$ is often regarded as a requirement of programs which implement Newton's method. However, $F^{(1)}$ is often complicated and might not be easy to specify if F is produced by a lengthy subroutine rather than by the use of a simple formula. In that case the use of a finite difference simulation of the derivative is useful, since it only involves the ability to calculate F. It has the disadvantage of requiring two calls to the F subroutine in estimating $F^{(1)}$, but the use of explicit F and $F^{(1)}$ would also require two function evaluations.

The forward simulation of the derivative leads to the formulae

$$SH = DE*F1/(F1 - F2) \tag{2.2}$$

$$E' = E + SH \tag{2.3}$$

with

$$F1 = F(E), \qquad F2 = F(E + DE) \tag{2.4}$$

DE is a fixed small increment (usually IE − 3 is satisfactory). Although some authors have expressed doubts about the use of this finite difference approach, the present author has always found it to be reliable for a wide variety of problems. The point about the expression (2.2) is that it does not attempt to give a very accurate $F^{(1)}$ value, since it is only a forward difference expression. Nevertheless, it produces the correct roots, since the shift SH reduces to zero when F(E) is zero. When (as in FIDIF or RADIAL) an exact zero value of F is not found because of the large value of $F^{(1)}$ at the root, the expression in (2.2) still leads to a zero shift.

Breaking the equation up into one which gives SH and one which adds it to E is a helpful device for computation. It permits, if desired, the inclusion of extra lines which test SH and reduce it in magnitude if necessary. If Newton's method is applied at an E value between two simple roots there is a danger that the test value will be near the extremum of F(E) (which must exist between the roots). The resulting calculated shift SH will be very large, producing a jump to a far-distant value of E. It is accordingly sensible to inspect SH for such anomalous behaviour before using it to revise the current estimate of the root. This is particularly important when NEWTON is used on its own, without the preliminary action of ROOTSCAN to narrow down the search region. The difficulty is, of course, inherent in Newton's method and is not due to the use of a forward difference formula to estimate the derivative. A further advantage of isolating SH before applying it is that a criterion such as $|SH| < 2E-8$ can be used to stop the iterations when the process has converged to the root.

SECANT

As correctly pointed out by Ralston and Rabinowitz (1978) the secant method is an old but neglected method which is suitable for use on computers. It has some similarities with the method used in NEWTON, but estimates the derivative $F^{(1)}$ by using the two latest E and F values directly, instead of imposing a fixed shift on the E value. The relevant formulae are thus

$$SH = (E - ES)*FS/(FS - F) \tag{2.5}$$

$$E' = E + SH \tag{2.6}$$

where ES and FS (*stored* values) are the values of E and F(E) from the *previous* iteration. Once the secant process is started it only requires one function evaluation per iteration, rather than two. It usually finds a root more quickly than NEWTON, although requiring a greater number of iterations. All of the tests on SH described for NEWTON can also be applied within SECANT.

ZIGZAG

Most of the functions produced by the subroutines in this book have simple zeros. However, the matrix program FOLDER, which uses real arithmetic, can be modified to treat Hermitian matrices, and then gives rise to a function which has only double zeros. As discussed by Channabasappa (1979), Newton's method will work for an m-fold root if the traditional formula is modified to the form

$$E' = E - mF(E)/F^{(1)}(E) \tag{2.7}$$

For a double root (with $m = 2$) this modified formula gives an iterative process with second-order convergence. The formula with $m = 1$ will still work for a double root, but will only give linear convergence; the error is halved at each iteration, although an Aitken extrapolation would restore the second-order convergence.

To find a double root Newton's method (or its finite difference approximation NEWTON) would have to be used on its own, since the preceding module ROOTSCAN would not detect a sign change and so would not pass control over to the NEWTON module. ZIGZAG represents another approach, in which the root-finding process is more closely integrated with the scanning process. It is based on the idea that on passing through a root of any multiplicity the ratio $R = F(E + DE)/F(E)$

will become greater than one for a sufficiently small DE. From the R value the position of the zero is estimated by extrapolating back linearly, using R for single roots and $R^{1/2}$ for double roots. From this improved E value the scan is continued with DE replaced by $-DE/8$. When R exceeds one again the process is repeated, and so on until E has converged. The successive estimates zigzag towards the root in an oscillating convergence. If the extrapolation based on R is omitted, so that only the assignment $DE := -DE/8$ is used when $R > 1$, the process becomes a direct search for the root and still converges. Hyslop (1972) has pointed out that direct search methods for an extremum (which is what a double root is) can lead to loss of accuracy. The method with extrapolation is not too subject to this difficulty and also converges much more speedily.

For the case of a double root ($I = 2$) the possibility arises that the test values E1 and E2 might be on opposite sides of the root and still give $R > 1$. In that case the correct extrapolation formula would give a shift of $DE/(1 + R^{1/2})$ instead of $DE/(1 - R^{1/2})$. A program which tests to decide which situation holds can be written, but this added detail was found to be hardly more effective than simply reversing the sign of DE at each iteration and using the fixed formula $DE/(1 - R^{1/2})$ for the shift throughout.

2.4 A note on cubic equations

The root-finding methods used here employ only real arithmetic and search only for real roots. They are *not* rendered inoperative if the function has complex roots as well as real ones, since (for a continuous function) it will always be the case that changes of sign along the real axis indicate the location of real roots. Indeed, it is sometimes an advantage to be able to see only real roots; for example, the Hill determinant method (see SEROSC) can produce many complex roots as well as the real ones which represent energy levels.

In the case of a cubic equation with real coefficients a special simplification occurs in the algebra; this makes possible the determination of the two complex roots (when they occur) by real arithmetic. For the equation

$$Ax^3 + Bx^2 + Cx + D = 0 \tag{2.8}$$

the sum and product of the roots are given by

$$\sum = -B/A \qquad \prod = C/A \tag{2.9}$$

and are known. If the real root R, which always exists, is located by any of the root-finding routines, then the assumption that the remaining two roots take the form $X \pm Y$ leads to the following algebraic results:

$$X = \left(\sum - R\right)/2 \tag{2.10}$$

$$Y^2 = X^2 - \prod/R = G. \tag{2.11}$$

If $G > 0$, then the roots are $X \pm G^{1/2}$. If $G < 0$ then the roots are $X \pm \mathrm{i}|G|^{1/2}$. Thus the two remaining roots of the cubic equation are computable immediately when a real root has been located. A short routine after the root-finding module can be used to produce the extra roots, as pointed out by Killingbeck (1985a).

2.5 Brief program descriptions and programs

The programs are intended to be used as modules within larger programs, but have been provided here with a few lines of removable program 'fore and aft', so that their operation can be checked.

ROOTSCAN

Line 380 keeps the scan going until a sign change in F(E) is observed. In the version shown here a result R < 0 leads to a stop with a 'RETURN without GOSUB' error. In rare cases F1 can be computed as zero (i.e. as less than the underflow level), so that a new run with different DE is needed. An alternative is to have an extra line such as

```
355 IF F1=0 THEN LET R=IE20:GO TO 370
```

Activating line 370 produces a slow zigzag search for a real root.

Line 390 works out an interpolated E value from the results at E1 and E2, ready for passing to the root-finding module. This line is on a border between modules and so can be incorporated in the root-finding module if desired.

NEWTON

Line 400 sets the E increment EI and the limit LIM on the allowed shift in the root estimate.

Lines 450 and 460 (remmed out here) are typical lines useful in controlling the shift and in checking for convergence of the iterative process.

SECANT

The program uses the current values E and F and the previous (stored) values ES and FS in estimating the position of the root. Lines 460 and 470 can be used in the same way as for NEWTON.

ZIGZAG

As displayed here the program allows a choice of the method used (M = 1 for direct search, M = 2 for extrapolation plus search) and of the multiplicity I of the root being sought. (The choice of I is particularly intended to be used when double roots are *known* to be involved.)

Line 290 stores DE0, since DE is reduced during the iteration.
Lines 355 to 365 accomplish the extrapolation, and are not used when M is 1. The power of R taken in line 355 is appropriate for the multiplicity of the root.

Although the program segments are shown separately, it is clear that the program block most commonly needed within larger programs is one consisting of ROOTSCAN plus one of the root-finder modules. A ROOTSCAN–SECANT or ROOTSCAN–NEWTON program block can be recorded as a separate entity and then inserted into another program by merging input. Alternatively, it can be fed into the microcomputer as the starting section, with other program lines being typed in around it.

```
 200 INPUT "E0,DE";E0,DE
 297 REM  **********************************
 298 REM  ROOTSCAN
 299 REM  **********************************
 300 LET E=E0
 310 GO SUB 1000: PRINT E,F
 320 LET F1=F: LET E1=E
 330 LET E=E+DE
 340 GO SUB 1000: PRINT E,F
 350 LET F2=F: LET E2=E
 360 LET R=F2/F1
 370 REM IF R>1 THEN LET DE=-DE/4: GO TO 310
 380 IF R>0 THEN GO TO 320
 389 REM  **********************************
 390 LET E=E2-DE+DE/(1-R)
 391 REM  **********************************
1000 LET F=(E-1)*(E-1)+.1
1010 RETURN
```

```
 300 INPUT "E";E
 310 LET DE=.4
 397 REM  **********************************
 398 REM  NEWTON
 399 REM  **********************************
 400 LET EI=1E-3: LET LIM=ABS (DE/4)
 409 REM  **********************************
 410 GO SUB 1000: PRINT E,F
 420 LET ES=E: LET FS=F
 430 LET E=E+EI: GO SUB 1000
 440 LET SH=EI*FS/(FS-F)
```

```
 450 REM IF ABS SH>LIM THEN LET SH=LIM*SGN SH
 460 REM IF ABS SH<2E−8 THEN GO TO 480
 470 LET E=E−EI+SH: GO TO 410
 477 REM  ********************************
 478 REM  BACK TO ROOTSCAN
 479 REM  ********************************
 480 LET E=E2: GO TO 310
 481 REM  ********************************
1000 LET F=(E−1)*(E−1)×.1
1010 RETURN
```

```
 300 INPUT "E";E
 310 LET DE=.1
 320 GO SUB 1000
 330 LET E1=E: LET F1=F
 340 LET E=E+DE: GO SUB 1000
 350 LET E2=E: LET F2=F
 360 LET R=F2/F1
 397 REM  ********************************
 398 REM  SECANT
 399 REM  ********************************
 400 LET E=E2−DE+DE/(1−R)
 410 LET LIM=ABS (DE/4)
 420 LET FS=F2: LET ES=E2
 430 GO SUB 1000: PRINT E,F
 440 LET SH=F*(E−ES)/(FS−F)
 450 LET ES=E: LET FS=F
 460 REM IF ABS (SH/E)<2E−8 THEN GO TO 490
 470 REM IF ABS SH>LIM THEN LET SH=LIM*SGN SH
 480 LET E=E+SH: GO TO 430
 487 REM  ********************************
 488 REM  RETURN TO ROOTSCAN
 489 REM  ********************************
 490 LET E=E2: GO TO 310
 491 REM  ********************************
1000 LET F=(E−1)*(E−2)×.1
1010 RETURN
```

```
 280 INPUT "M,I";M,I
 287 REM  ********************************
 288 REM  ZIGZAG
 289 REM  ********************************
 290 INPUT "E0,DE0";E0,DE0
 300 LET E=E0: LET DE=DE0
 305 GO SUB 1000: PRINT E,F
 310 LET E1=E: LET F1=F
 320 LET E=E+DE
 325 GO SUB 1000
 330 LET E2=E: LET F2=F
```

```
335 IF F1=0 THEN LET R=1E20: GO TO 345
340 LET R=F2/F1
345 IF R<1 THEN GO TO 310
346 REM  *******************************
350 IF M=1 THEN GO TO 370
354 REM  *******************************
355 LET R=R ↑ (1/I)
360 LET SH=DE/(1-R)
365 LET E=E1+SH
369 REM  *******************************
370 LET DE=-DE/8
375 IF ABS ((E1-E)/E)<2E-8 THEN PRINT E: STOP
380 GO TO 305
381 REM  *******************************
1000 LET A=3: LET B=.1
1010 LET F=(E-1)*(E-A)+B
1020 RETURN
```

2.6 Extrema and roots

The values of a function in the neighbourhood of a particular point $x = x_0$ can be represented by a Taylor series expansion; the leading terms give the approximation

$$F(x_0 + h) = F(x_0) + hF^{(1)}(x_0) + (h^2/2)F^{(2)}(x_0). \quad (2.12)$$

To find a zero near to x_0 we set $F(x_0 + h)$ equal to zero and obtain a quadratic equation for h, which can be written as

$$h = -[F(x_0) + (h^2/2)F^{(2)}(x_0)]/F^{(1)}(x_0) \quad (2.13)$$

if the derivative $F^{(1)}(x_0)$ is not zero. The first approximation to the solution of (2.13) is obtained by neglecting the h^2 term, which gives just Newton's formula. To use this first h value in (2.13) again to get a corrected shift is possible, but requires the value of $F^{(2)}(x_0)$ and so is usually not as efficient as simply iterating using the standard Newton formula. In NEWTON the derivative $F^{(1)}$ is approximated by a forward difference formula. This means that only two x values are used to estimate h. Most simple root finders use either one or two function evaluations per iteration. If we attempt to find a point near x_0 at which the derivative $F^{(1)}$ is zero, we obtain the result

$$h = -F^{(1)}(x_0)/F^{(2)}(x_0) \quad (2.14)$$

which is Newton's formula for the case in which a zero of $F^{(1)}$ (rather than F) is required. To obtain an estimate of $F^{(2)}$ from function values requires *three* function values. If the three test points are at x_0, $x_0 \pm \varepsilon$, we can represent the derivatives at x_0 by central difference estimates. Newton's formula for the zeros of F then gives (in an obvious notation)

$$h = -2\varepsilon F_2/(F_3 - F_1) \tag{2.15}$$

which requires the values of F at three points and is less speedy than the use of the forward difference formula. Newton's formula (2.14) for the zeros of $F^{(1)}$ becomes

$$h = -(\varepsilon/2)(F_3 - F_1)/(F_1 + F_3 - 2F_2) \tag{2.16}$$

and could be used to find stationary points of F together with an estimate of $F^{(2)}$, which would indicate the nature of the stationary point.

If $F_3 = F_1$ then the h value obtained from (2.16) is zero, which indicates that x_3 and x_1 exactly straddle the stationary point at x_2. This observation indicates a way to use only *two* points at a time, by using the function $F(x + \varepsilon) - F(x - \varepsilon)$ in a root-finding program. ε should be small, but the dependence of the result on the ε value can be tested empirically. In a direct search method it is necessary for $f(x_0)$ and $f(x_0 + \varepsilon)$ to be computed as different numbers in order for x_0 to be clearly identified as an extremum. Since the difference between these function values varies as ε^2 near x_0, there will be a band of x values around x_0 within which $f(x_0)$ and $f(x_0 + \varepsilon)$ are indistinguishable with a given computer precision; this leads to a necessary degree of uncertainty in the value of x_0 as found by a direct search method. Hyslop (1972) gave several examples of this, but the effect can be demonstrated even for a simple polynomial.

Using the simple program SEARCH (given later), the function

$$F(x) = x(x - 2)(x - 5) \tag{2.17}$$

can be computed for increasing x to observe the positions of the extrema. It is found that, on a typical microcomputer, F is computed to have the common value 4.060 672 6 for all x values in the interval between 0.880 28 and 0.880 45. The use of a simple search program such as SEARCH produces a result for the position of the maximum which depends on the starting x value, the starting D value and also the factor by which D is divided when its sign is reversed. The best that can be concluded is that the maximum is probably somewhere between 0.880 35 and 0.880 37. However, the function $F(x + \varepsilon) - F(x - \varepsilon)$ can be used as the subroutine function in the root-finder ZIGZAG. If the simple direct search version (without extrapolation) is used, then we are keeping the root finding as simple as possible and concentrating on the change in the function under consideration. The results are remarkable: with $\varepsilon = 0.02$ roots at 0.880 412 91 and 3.786 253 7 are found, while with $\varepsilon = 0.01$ these roots move to 0.880 378 48 and 3.786 288 1. These represent the roots of the function

$$F(x + \varepsilon) - F(x - \varepsilon) = 2\varepsilon F^{(1)}(x) + (\varepsilon^3/3)F^{(3)}(x) + \ldots \tag{2.18}$$

and can be found accurately since the ε value is quite large enough for $F(x + \varepsilon)$ to differ markedly from $F(x)$. The result (2.18) indicates that it should be possible to obtain an improved result by a process of extrapolation. The leading error in the root position is of ε^2 type, and a 4 : − 1 weighted average of the computed results yields the corrected roots 0.880 367 00 and 3.786 299 6, which agree with the analytical results $(7 \pm 19^{1/2})/3$.

If the function $F(x)$ being studied is given by a complicated subroutine, then the subroutine will have to be traversed twice to compute $F(x+\varepsilon) - F(x-\varepsilon)$. This can be accomplished by imbedding the function subroutine inside a loop which is traversed twice (with appropriate x values) so that the required difference is finally returned as the function value. This approach of inserting a few extra lines into the function module leaves the root-finding module unchanged. Alternatively, a special extremum-finding module can be written to perform the task of calling the function subroutine twice. The method of using root finding together with extrapolation to locate extrema also serves to find the maximum of the function

$$F(x) = \exp(-x) - \exp(-2x) \tag{2.19}$$

treated by Hyslop (1972). The method used here produces the result 0.693 147 2 when the ε values 0.04 and 0.02 are used. This result agrees with the correct result ln 2; a direct search for the maximum is only able to locate it as being between 0.693 01 and 0.693 29, since all x values in this range produce a computed function value of 0.25 (on the particular microcomputer used for the test).

2.7 The programs SEARCH and MAXMIN

The program SEARCH as listed here will search for a maximum of the function evaluated in the function subroutine at line 1000. An obvious change in line 350 will make it search for a minimum. The program MAXMIN looks for extrema by applying the simple algebra discussed above, the three sample points being X, X ± D. If RF in lines 110 and 400 is 1, then MAXMIN reproduces the results at fixed D (0.02, 0.01, etc) which were quoted above as alternatively coming from a modification of ZIGZAG. If RF is set equal to 1.1, then the output X values will move towards a fairly good estimate of the extremum position, until rounding errors produce fluctuations. However, the extrapolation process based on fixed D values is more reliable, and is actually an example of the Richardson extrapolation process which is applied repeatedly throughout this book.

```
297 REM ********************************
298 REM  SEARCH
299 REM ********************************
```

```
300 INPUT "X,D";X,D
309 REM ********************************
310 GO SUB 1000: PRINT X,F
320 LET F1=F: LET X=X+D
330 GO SUB 1000: PRINT X,F
340 LET F2=F
350 IF F2>F1 THEN GO TO 320
360 LET D=-D/4: GO TO 310
361 REM ********************************
1000 LET F=X*(X-2)*(X-5)
1010 RETURN
```

```
 97 REM ********************************
 98 REM  MAXMIN
 99 REM ********************************
 100 INPUT "X,D";X,D
 110 INPUT "RF";RF
 199 REM ********************************
 200 LET X2=X
 210 GO SUB 1000: LET F2=F
 220 LET X=X+D: LET X3=X
 230 GO SUB 1000: LET F3=F
 240 LET X=X2-D: LET X1=X
 250 GO SUB 1000: LET F1=F
 299 REM ********************************
 300 LET NUM=(F1-F3)*D/2
 310 LET DEN=F1+F3-2*F2
 320 LET SH=NUM/DEN
 330 LET X=X2+SH
 340 PRINT X
 399 REM ********************************
 400 LET D=D/RF: GO TO 200
 997 REM ********************************
 998 REM  FUNCTION SUBROUTINE
 999 REM ********************************
1000 LET F=X*(X-2)*(X-5)
1005 REM LET F=EXP (-X)-EXP (-2*X)
1010 RETURN
1011 REM ********************************
```

2.8 Expectation values. The integral approach

The traditional way to define an expectation value in quantum mechanics is by means of an integral involving the normalized wavefunction. If the wavefunction ψ produced by a calculation is not normalized, then a ratio of two integrals is required. For example the expectation value $\langle x^2 \rangle$ of x^2 for a one-dimensional problem is found from the equation

$$\langle x^2 \rangle \int \psi^2 \, dx = \int \psi^2 x^2 \, dx \tag{2.20}$$

where the limits of integration are the boundary points at which $\psi(x) = 0$ (for the case of Dirichlet boundary conditions). The definition (2.20) can be used computationally, but requires both the evaluation of $\psi(x)$ throughout the region considered and the estimation of the two integrals using some method such as the trapezoidal rule. We can immediately see two cases in which this would cause extra difficulties. The first case is that of an unstable finite difference shooting method which produces accurate energy eigenvalues but does not give a reliable wavefunction over the outer parts of the region of integration. The second case is that of a power series method which works out the wavefunction on the boundary of the region considered and not throughout the entire region. The programs RADIAL and SEROSC of this book exemplify these methods.

2.9 The energy approach

To obtain expectation values from methods such as the two quoted above we note that they both belong to the root-finding family of methods. In fact they both use the condition $\psi(E, L) = 0$ for a fixed boundary value L, so that the eigenvalue calculation involves finding the roots of some function of E. To exploit this fact we use a result of first-order perturbation theory: if a small perturbing term εx^2 is added to the potential in the Schrödinger equation then each eigenvalue E_n is shifted by an amount $\varepsilon\langle x^2 \rangle$, where the expectation value is taken with respect to the unperturbed eigenfunction ψ_n associated with E_n. This fact provides us with an alternative way of defining an expectation value for an eigenfunction; the expectation value becomes in effect an energy response coefficient. For example, $\langle x^2 \rangle$ for an eigenfunction can be defined as the rate of change of the eigenvalue with respect to the coefficient of x^2 in the potential function.

There are two obvious ways to implement this energy-based definition computationally. The first way is to use a central difference formula to define the derivative, obtaining the result (for the nth state)

$$\langle x^2 \rangle_n = \operatorname*{Lt}_{\varepsilon \to 0} [E_n(V + \varepsilon x^2) - E_n(V - \varepsilon x^2)]/2\varepsilon. \tag{2.21}$$

This approach is applicable with several methods of eigenvalue calculation, but necessarily involves a numerical differentiation and the performance of at least two separate calculations of the nth eigenvalue E_n. The second way to apply the response coefficient definition of an expectation value is to calculate the required derivative directly and exactly (apart from rounding error) within the same computation which yields the wavefunction. Killingbeck (1985c) pointed out that this can be done for those methods which involve root finding, so that each eigenvalue is a root of a function $F(E)$. Suppose, for example, that we wish to calculate $\langle x^2 \rangle$. We write the function in the form $F(E, \varepsilon)$, to indicate that the value of the function F will depend on E and also on the coefficient ε, where εx^2 is a term in the potential function. (F also depends on other coefficients in the potential, of course, but we are not going to vary them.) When ε changes slightly to $\varepsilon + \mathrm{d}\varepsilon$, then F will change slightly and the original root E will move to $E + \mathrm{d}E$ in such a way as to ensure that F remains zero. Thus, both before and after varying ε, the value of F is zero. The fact that F does not change means that we must have

$$0 = \mathrm{d}\varepsilon(\partial F/\partial \varepsilon)_E + \mathrm{d}E(\partial F/\partial E)_\varepsilon \qquad (2.22)$$

from which we conclude that

$$\langle x^2 \rangle = \mathrm{d}E/\mathrm{d}\varepsilon = -(\partial F/\partial \varepsilon)_E/(\partial F/\partial E)_\varepsilon. \qquad (2.23)$$

The beauty of this result is that it is usually possible to compute both of the derivatives on the right by making some modification of the existing subroutine which works out the function F. Further, if the value of $\partial F/\partial E$ is computed, it then becomes possible to use it directly in Newton's method for locating the roots of the function F, without using the finite difference approach contained in the root-finding program NEWTON. When the program for computing F is modified in this way the result is a program in which the root-finder module can be simplified, since the function subroutine gives both F and $\partial F/\partial E$ at each call.

2.10 Calculating local quantities

In a finite difference calculation (e.g. FIDIF) it is possible to make an amusing but useful extension of the methods outlined above. The expectation value of a Dirac delta function for an eigenfunction would take the form

$$\int \psi^2(x)\delta(x - y)\,\mathrm{d}x = \psi^2(y) \qquad (2.24)$$

and so would yield the value of ψ^2 at the point y. Here ψ is the *normalized* wavefunction, even though the calculation involves no specific integration to obtain a normalization constant, and need not even store any of the wavefunction values. Similarly, if we evaluate the expectation value of a function which has the value 1 between $x = A$ and $x = B$, and 0 elsewhere, then we obtain the quantum mechanical probability value associated with the interval $A \leqslant x \leqslant B$ (i.e. the integral of ψ^2 between A and B). These two approaches have been used in a study of double-well potentials (Killingbeck 1988b) to monitor how slight asymmetry in the potential can cause considerable asymmetry in the probability distribution.

To simulate a delta function situated at the gridpoint $x = y$ the use of a Boolean function simplifies matters. For example, in a finite difference calculation using step length H the potential

$$(\mathrm{X} < \mathrm{Y} + \mathrm{H}/2) * (\mathrm{X} > \mathrm{Y} - \mathrm{H}/2)/\mathrm{H} \tag{2.25}$$

will suffice, as shown by Killingbeck (1988b). The simple choice (X = Y)/H is risky; if the program adds a floating point H to a floating point X to advance a step at a time along the x axis, then it might miss the exact point X = Y because of a slight rounding error. For certain special X values this problem can be avoided by using integer variables. If a counting integer N (1 to NS) is used to count the steps, with NS steps in the full region of integration, the Boolean function (N = NS/2)/H would reliably simulate a delta function at the midpoint of the region.

The discussion of this section has indicated the special methods which are made possible when the root-finding approach (which is of wide applicability) is adopted in dealing with the Schrödinger equation of quantum mechanics. The result is a kind of quantum mechanics in which the computational emphasis is on eigenvalues rather than eigenfunctions. The programs SERAT, SEROSC, RADIAL, FIDIF and HYPOSC exemplify this in varying degrees, but it is important to remember that the underlying *algebraic* analysis played a central role in making that computational approach possible.

3 The Richardson extrapolation method

PROGRAMS

RICH, ROMBERG.

3.1 General introduction

This chapter sets out the main techniques involved when the problems of differential and integral calculus are simulated by finite difference formalisms for use on a computer. The method of Richardson extrapolation is explained as a way of taking the limit $h \to 0$ of the calculus by numerical means, and examples involving differentiation and quadrature are given. Chapter 2 gave an application from the theory of maxima and chapter 8 gives examples from eigenvalue calculations for the Schrödinger differential equation. Section 3.5 comments on the differences and similarities between the Richardson and rational (Padé) methods of extrapolation, and the chapter includes the program RICH, which can be used to perform Richardson extrapolation in conjunction with various other programs. The program ROMBERG for numerical quadrature is introduced in a somewhat unorthodox manner; both the method of calculating the coefficients A_2 and A_4 and the establishment of three-point rules from two-point rules were specially devised to illustrate how the use of the appropriate mathematical approach can lead to computational advantages.

Richardson extrapolation

3.2 Microcomputer calculus

Several of the calculations described in this book involve the use of a finite difference approximation to some operation of the calculus.

Nowadays there exist several useful programs (e.g. MUMATH, MACSYMA) which can perform algebraic manipulations, including differentiation and integration, on functions which are given explicitly in terms of standard functions of exponential and trigonometric type. Some of these programs have been sufficiently condensed to work on modern microcomputers, although the more traditional numerical approaches which are treated in this book can be used on microcomputers of much less memory capacity than that required by symbolic manipulation programs.

For the most part our numerical calculations are based on simple formulae which give numerical estimates of a first or second derivative at a point, or of the area contained within a strip of width h under the graph of a function. Such formulae are used, for example, in estimating the gradient of a function for use in Newton's method for finding the roots of an equation (NEWTON), in simulating the operation of the kinetic energy operator in a finite difference approach to the Schrödinger differential equation (FIDIF) and in estimating the numerical value of a definite integral (ROMBERG). Analysis of each formula by means of a Taylor series expansion indicates the kind of discretization error which it contains, and the principal idea of what is usually called Richardson extrapolation is to correct for this error in order to obtain the correct 'calculus' result which would be obtained in the limit $h \to 0$, where h is the strip width or step length actually used in the finite difference calculation. This limit usually has to be estimated theoretically, since an attempt to approach it directly by using smaller and smaller h values in actual computations would give longer and longer running times and also increased rounding errors due to the increasing number of arithmetic operations involved. Indeed, the increasing rounding error would actually mean that we were aiming at a moving target!

3.3 An example. Numerical differentiation

We can illustrate the idea behind Richardson extrapolation by following Rutishauser (1963), who applied it to numerical differentiation. In estimating the derivative of a function at a point x we usually employ the central difference quantity $y(x + h) - y(x - h)$, for which we can write down the result

$$y(x + h) - y(x - h) = 2hf^{(1)}(x) + (h^3/3)f^{(3)}(x) \dots \quad (3.1)$$

which is obtained by starting from the Taylor series for $y(x \pm h)$. Dividing throughout by $2h$ gives the result

$$[y(x + h) - y(x - h)]/2h = f^{(1)}(x) + (h^2/6)f^{(3)}(x) + \dots \quad (3.2)$$

which indicates that the traditional central difference formula for the derivative has a leading error of h^2 type, with higher-order error terms of type h^4, h^6, etc. That the formal series for the error involves only even powers of h is due to the symmetric form of the central difference expression; this property is shared by most of the formulae commonly used in numerical quadrature or in the numerical solution of differential equations.

We would, of course, have anticipated that taking the limit $h \to 0$ gives the exact derivative, but we now know something about the *law of approach* to the correct value. There is an obvious way to exploit the equation (3.2). If we write it down again using the increment $h/2$ instead of h, and then regard the resulting two equations as a pair of simultaneous equations for the quantities $f^{(1)}(x)$ and $f^{(3)}(x)$, with higher-order terms neglected, then we find

$$f^{(1)}(x) = [4G(h/2, x) - G(h, x)]/3 \tag{3.3}$$

$$f^{(3)}(x) = [8G(h, x) - 8G(h/2, x)]/h^2 \tag{3.4}$$

where we have introduced the symbol $G(h, x)$ (G for gradient) to denote the central difference estimate of $f^{(1)}(x)$ using increment h. By checking the effect of the higher-order terms in equation (3.1) we find that the expression (3.3) for $f^{(1)}(x)$ now has a leading error term of h^4 type (instead of the original h^2 type), while the expression (3.4) for $f^{(3)}$ has a leading error term of h^2 type. If the estimates $G(h, x)$ and $G(h/2, x)$ are computed numerically in the course of some calculation, then we can estimate $f^{(3)}(x)$ using (3.4). More importantly, we can improve the estimates $G(h, x)$ and $G(h/2, x)$ for $f^{(1)}(x)$ by using (3.3), which is the most simple and commonly used example of a Richardson extrapolation formula. Rutishauser (1963) showed how effective this approach can be in calculating the derivative of a smooth function. Killingbeck (1985b) gave the example of the function $(1 + x^2)^{-1}$ at $x = 2$, where $f^{(1)}(x)$ equals -0.16 exactly. The numerical estimates $G(0.2, 2)$ and $G(0.1, 2)$ are $-0.161\,540\,45$ and $-0.160\,384\,25$, respectively, to eight significant digits. The two estimates agree in the first two significant digits only, but application of (3.3) gives the much improved estimate $-0.159\,998\,85$.

3.4 General applications

Although we have used the example of numerical differentiation here (actually the most notoriously dangerous operation in numerical calculus), suitably modified forms of equation (3.3) and its higher-order partners can be used in numerical quadrature (ROMBERG) and in finite difference eigenvalue calculations (FIDIF). The sections of the book dealing with those topics contain discussions of the way in which Richardson extrapolation can be used in each particular case. The original work of Richardson and Gaunt (1927) was carried out in the course of the numerical solution of some differential equations, and was called 'the deferred approach to the limit'. The application of the ideas to ordinary numerical quadrature is usually attributed to Romberg (1955), who specifically used the halving sequence $h, h/2, h/4$, etc, although the principles which he applied can be used for any sequence of decreasing strip widths. In general, since the extrapolation method involved can be applied to several different numerical calculations of differential and integral calculus, the most effective way to proceed is to have a specific subroutine which can carry out the extrapolation process on the output values for h, $h/2$, etc, which are produced by the main program. These values might be estimates of a derivative, an integral or an eigenvalue; all that is required is that for sufficiently small h they differ from their 'true' values (in the limit $h \to 0$) by an error term which can be represented as a series in terms of increasing powers of h.

3.5 Richardson versus Padé

The most common error series encountered are those involving even integer powers of h, usually with an h^2 or h^4 term as the leading one. However, there are cases in which terms with fractional powers of h or even terms involving $\log h$ can occur (Fox 1967). This means that a preliminary theoretical analysis may be necessary to decide upon the best way to apply Richardson extrapolation. For the case in which non-integer powers of h occur in the formal error series, a change of variable can sometimes convert the series to the normal form involving powers of h^2 (Killingbeck 1985a), or the application of a Padé approximant approach (using, for example, the subroutine WYNN) can be effective even without detailed knowledge of the powers of h which appear in the error series. It is usually possible to estimate the first few powers of h in the error series by numerical calculation; indeed, if these powers of h are already known from theory, one of the detailed tests which can be applied to a program is to check that it does actually give reasonable estimates for them. For example (to quote from the author's experience of student programming assignments), a program which supposedly applies the midpoint integration rule to estimate the value of a definite integral should (in theory) give a numerical result with a leading discretization error of h^2 form. If the error actually seems to be roughly proportional to h as h is decreased, a likely culprit is the program loop which counts the strips. If it counts one too many it adds on an extra strip at the end, thus introducing an extra error of approximate magnitude $hf(U)$, where U is the upper limit of the integration. We should note that in principle the Padé approximant approach would work correctly even with such an error present, since it would allow for the h-type error term and still produce a good estimate for the integral. Even with the inclusion of the extra strip, we still obtain the integral up to the intended limit U in the limit $h \to 0$, since the contribution of the unwanted strip tends to zero in that limit.

In cases where the form of the error series is *known*, either from theory or from computational experiment on test cases, Richardson extrapolation can be carried out efficiently using that knowledge. If we simply suppose that an error series in powers of h *exists*, but do not assume any knowledge of the powers involved, then the Padé approximant analysis can be applied (under conditions specified more fully in the section associated with the program WYNN). To apply the Padé approximant analysis, however, requires more values of the quantity concerned $(G(h), I(h), E(h), \text{etc})$ than are required for Richardson extrapolation. To illustrate this we recall that the two central difference estimates $G(0.2, 2)$ and $G(0.1, 2)$ for the derivative of $(1+x^2)^{-1}$ at $x = 2$ produced a much improved estimate of the derivative when we based our Richardson extrapolation on the knowledge that the leading error is of h^2 type. Without that knowledge, we need more values of G to make any progress in general. If we add the new computed result $G(0.4, 2) = -0.166\,212\,30$, we can apply the Padé approximant approach here by remembering that for a case such as this that approach gives a lowest-order result which (as explained in the WYNN section) is obtained by regarding the three G values just as though they contained errors which are the successive terms of a geometric progression. We have, for example, for *any* small h value, the supposed law

$$G(h, 2) = f^{(1)}(2) + Ah^N \tag{3.5}$$

where we consider only the dominant error term and regard the numbers A and N as unknown. Writing down the corresponding equations for $h/2$ and $h/4$ gives the results

$$\begin{aligned} G(h, 2) &= f^{(1)}(2) + \varepsilon \\ G(h/2, 2) &= f^{(1)}(2) + \varepsilon k \\ G(h/4, 2) &= f^{(1)}(2) + \varepsilon k^2 \end{aligned} \tag{3.6}$$

with $\varepsilon = Ah^N$ and $k = (1/2)^N$, thus showing that the error terms do indeed behave like the terms of a geometric progression for the special case in which each strip width used is a fixed fraction of the preceding one. Algebraic manipulation of the three equations above gives the result

$$G(h/4, 2) - G(h/2, 2) = k[G(h/2, 2) - G(h, 2)] \tag{3.7}$$

from which the common ratio k can be found. Our three quoted numerical values for G give the result $k = 0.2474822$. If we wish to take this k value as indicating that the unknown index N is 2 (since then k would be 1/4) we can proceed to combine the results for 0.2 and 0.1 to obtain the same result as before, $f^{(1)} = -0.15999885$. Applying the same h^2 extrapolation to the results for 0.4 and 0.2 would give us the new estimate $f^{(1)} = -0.15998317$. If we now suppose that the next term in the error series is of h^4 type (which we already established in connection with equation (3.3)) then we can quickly establish algebraically that these two 'improved' $f^{(1)}$ estimates should have h^4 error contributions which have the ratio 16 : 1. To eliminate that error contribution we should take a weighted average of the two results in the proportions 16/15: −1/15. This yields the final best estimate $f^{(1)} = -0.15999990$ from the Richardson extrapolation based on three G values instead of the two which were originally used. In performing this analysis, however, we have gone *beyond* the error terms appearing in the three equations. If we refuse to take this extra step then the best we can do is to put the calculated k value back into the equations in order to find $f^{(1)}(2)$ and ε. We can combine the first two equations, for example, to obtain the result

$$(k - 1)f^{(1)}(2) = kG(h, 2) - G(h/2, 2) \tag{3.8}$$

from which we find the estimate $f^{(1)}(2) = -0.16000401$, which is also the result obtained if we use the last two equations instead. The error is calculated to be $\varepsilon = -0.00620829$ using the first equation.

In the calculations described above, we have proceeded as though the exact answer $f^{(1)}(2)$ is unknown, since that most often *is* the case when we are tackling a new problem. If, however, we assume right from the start that we are doing a test calculation in which $f^{(1)}(2)$ is *known* to have the exact value −0.16, then we know directly the errors in the computed G values. We find, for example, that the quantities $G(0.2, 2)$ and $G(0.1, 2)$ differ from $f^{(1)}(2)$ by errors which are in the ratio 4.01 : 1. This strongly suggests a dominant term of h^2 type in the error series. The analogous error ratio for $G(0.4, 2)$ and $G(0.2, 2)$ is 4.03 : 1, which also suggests an h^2 term. Once the h^2 law has been confirmed on the test case it can then be used to help us in cases where the derivative is not known analytically but can only be estimated numerically. Some of the subroutine functions F(E) arising in this book fall into that category (the programs FOLDER and MOMOSC involve such functions). The simple example of numerical differentiation given above has been discussed at some length, since the methods used are also applicable to most problems in which finite difference formulae are used to simulate the operations of the calculus. What the discussion should have made clear is that to find the best approach in each case usually involves applying a mixture of theoretical insight and computational experiment, a procedure for which interactive computing is highly suitable.

3.6 The residual error term

The use of Richardson extrapolation or the Padé approximant approach in connection with computing results in the limit $h \to 0$ is based on the supposition that the discretization error can be represented by a series in powers of h, with the terms having rapidly decreasing magnitudes. In general, however, the formal error series is usually as asymptotic one, and a residual term should be added to it to produce the exact discretization error. This residual term has been investigated in the most detail for the case of numerical quadrature (Lyness and Ninham 1967), and the Euler–Maclaurin summation formula (Schwartz 1969) is a classical part of the relevant theory. The main point which emerges from detailed studies is that the residual term associated with the commonly used techniques falls off faster than any power of h as h tends to zero. This means that it is always possible to find h values which are sufficiently small for the residual term to be negligible (to the accuracy required) and yet sufficiently large for Richardson extrapolation to be much more effective than the crude method in which smaller and smaller h values are used until the results for h and $h/2$ agree to some specified number of digits.

3.7 The standard extrapolation formulae

For the case in which the error series is a standard one with terms in h^2, h^4, and so on, it is not difficult to formulate a simple recursive algorithm to compute the results of a Richardson extrapolation. If the results are to be set out in a table it is possible to use several different ways of labelling them. The obvious way is to label them like the elements of a matrix, so that the element $F(J, K)$ is the Jth element in the column K. Another approach is to let $F(J, K)$ denote the extrapolant which is based on the function values from $F(J)$ to $F(K)$ inclusive. This notation makes it clear at once how much information has been used in producing any given element of the array. The initial data values can be regarded in two different ways, either as functions of h or as functions of the number of strips N used in the integration process which yielded the results. To interpret results based on the strip width h in terms of a formula based on strip number N it suffices to use the reciprocal h^{-1} as the effective N for each data value.

To produce a specific formula we suppose that the data values have been set out in the form of a column, with the strip number N increasing down the column. We do *not* stipulate that the successive N values must have any fixed ratio (as Romberg did in his work on numerical quadrature). The case of four data values can be set out as shown below, with each data value $F(N)$ being given the double label $F(N, N)$.

$$\begin{array}{lllll}
N(1) & F(1,1) & & & \\
 & & F(1,2) & & \\
N(2) & F(2,2) & & F(1,3) & \\
 & & F(2,3) & & F(1,4). \\
N(3) & F(3,3) & & F(2,4) & \\
 & & F(3,4) & & \\
N(4) & F(4,4) & & &
\end{array}$$

The labelling is set up so as to indicate the range of data points which have been used in producing each Richardson extrapolant $F(J, K)$. When this standard layout is adopted the formulae which relate the values in the table take the simple form

$$F(J, K) = [RF(J+1, K) - F(J, K-1)]/(R-1) \qquad (3.9)$$

where

$$R = [N(K)/N(J)]^2. \qquad (3.10)$$

For the common case in which each $N(K)$ is half its predecessor the formulae give just the usual weighting factors which are powers of two, but are also applicable for the case of general $N(K)$. It is not even necessary for the $N(K)$ to be varying monotonically, although it *is* better to make $N(K)$ increase with K, so that the approximants down each column will then tend to vary systematically, making it easier to judge the degree of convergence of the extrapolation process.

3.8 The non-standard case

If the error series involves terms in $h^{1/2}$, $h^{3/2}$, etc, then the simple algorithm described above will need to be modified. We can suppose that the sequence of powers involved is *known* from some theoretical analysis or prior computational test, so that the indices (1/2, 3/2, etc) can be used as input data along with the $N(K)$ and $F(K)$ values. The extrapolation process for this more general case has been discussed by several authors. Håvie (1979) pointed out that the most simple case is that in which the $N(K)$ increase in a definite geometrical progression. In that case the formula (3.9) of the preceding section can still be applied. The only change necessary is to use the value ρ^I for R, where ρ is the common ratio for the geometric progression of $N(K)$ and I is the index $(1/2, 3/2, \text{etc})$ appropriate for the column of approximants being calculated. The formula (3.10) gives this result also for the case of a well defined ρ, but with $I = 2, 4, 6$, etc. The interesting fact is that the result extends to the case of arbitrary indices when it is suitably interpreted.

3.9 The general case. The program RICH

Even when the $N(K)$ do not form a geometric progression it is clear that the *first* column of approximants can be calculated using R values equal to $[N(K)/N(J)]^{I(1)}$. However, the calculation for later columns becomes more complicated. The direct way to perform the extrapolation then is to go back to first principles and solve the relevant equations sequentially, keeping track of all the multiplying factors required to eliminate the successive terms in the error series. The program RICH does this. It is necessarily more lengthy than the program needed to implement the simple formulae of the standard case, but it has the advantage that it produces as output either the numerical extrapolants *or* the numerical factors which are needed in forming the table of extrapolants. Given these numerical factors, it would be possible to carry out later calculations (for the same set of $N(K)$ and indices) using a pocket calculator, for example.

The program RICH is fairly self-explanatory, since it represents a translation into BASIC of the sequence of steps which would be carried out to eliminate the error terms one at a time. The assumption is that we have a set of quantities (called 'energies' here) which obey the law

$$E(K) = E + A[N(K)]^{-I(1)} + B[N(K)]^{-I(2)} + \ldots \qquad (3.11)$$

with the $N(K)$ and $I(K)$ known. The program carries out the elimination of A, B (and so on) successively by forming linear combinations of the equations in just the same way as it would be done in a hand calculation. The only complexity arises in deciding how to represent all the required extrapolants and multiplying factors as arrays, since the labelling must allow for us to move down and along the equations just as we would if they were written down on the page during a hand calculation.

The main points about RICH are as follows.

1. The formation of working copies of the strip numbers and data input values, so as to preserve the input data. A manual GO TO 2400 will allow a repetition of the analysis to get either the extrapolants or the table of coefficients without a second input of data.

2. By putting in the indices last we can try a new set of indices for the same input data, in case there is some doubt about their correct values.
3. Starting from the obvious set of coefficients based on the given indices (in line 2450) the calculation modifies the coefficients exactly in step with the calculated extrapolants (lines 2560 and 2580) and prints out either set of quantities in response to the choice made in line 2500. The display is put on the screen in columns separated by a space and is not stored in an array (although this modification is easy to add).
4. If the elimination between equations is done exactly, then various coefficients should be reduced to zero as the calculation proceeds. The program prints these coefficients as a check (line 2600), putting the word ZERO beside them.
5. The input for the E(K) (line 2200) is written for manual input, but a statement such as LET E(K)=T(K) could be used to copy over a T array from a preceding program. The strip numbers N(K) can also be set automatically for the standard case if INPUT C in line 2120 is changed to LET C=2*M.

```
1997 REM  **********************************
1998 REM  RICH
1999 REM  **********************************
2000 INPUT "NO.OF TERMS";Q
2010 DIM A(Q,Q): DIM E(Q)
2020 DIM B(Q): DIM S(Q)
2030 DIM C(Q): DIM I(Q)
2099 REM  **********************************
2100 PRINT "STRIP NOS,IN INCREASING ORDER"
2110    FOR M=1 TO Q
2120 PRINT M: INPUT C
2130 PRINT C: LET C(M)=C
2140    NEXT M
2199 REM  **********************************
2200 PRINT "ENERGIES"
2210    FOR M=1 TO Q
2220 PRINT M: INPUT B
2230 PRINT B: LET B(M)=B
2240    NEXT M
2299 REM  **********************************
2300 PRINT "INDICES"
2310    FOR M=1 TO Q-1
2320 PRINT M: INPUT I
2330 PRINT I: LET I(M)=I
2340    NEXT M
2399 REM  **********************************
2400 FOR M=1 TO Q: LET S(M)=C(M)
2410 LET E(M)=B(M)
```

```
2420 NEXT M
2429 REM  *********************************
2430 FOR M=1 TO Q
2440    FOR N=1 TO Q-1
2450 LET F=S(M) ↑ (I(N))
2460 LET A(M,N)=1/F
2470    NEXT N
2480 NEXT M
2499 REM  *********************************
2500 INPUT "0 FOR FACTORS,1 FOR ENERGIES";Z: CLS
2510 FOR N=1 TO Q-1
2520    FOR M=1 TO Q-N
2530 LET R=A(M,N)/A(M+1,N)
2540 IF Z=0 THEN PRINT R
2550    FOR J=N TO Q
2560 LET A(M,N)=(R*A(M+1,J)-A(M,J))/(R-1)
2570    NEXT J
2580 LET E(M)=(R*E(M+1)-E(M))/(R-1)
2590 IF Z<>0 THEN PRINT "      "; E(M)
2600 IF Z=0 THEN PRINT "ZERO";A(M,N)
2610    NEXT M
2620 PRINT ""
2630 NEXT N
2631 REM  *********************************
2640 PRINT "GOTO 2400 TO REPEAT"
2641 REM  *********************************
```

ROMBERG. Mathematical theory

3.10 Midpoint integration

The simple midpoint rule for numerical quadrature approximates the integral of a function $f(x)$ over a strip of width h by taking the product of h and the value of f at the midpoint of the strip. This rule is more accurate than the trapezoidal rule, which is more commonly treated in textbooks and which uses the product of h and the average of the f values at the ends of the strip. To simplify the theory of the midpoint rule it is best to take the origin of x at the midpoint of the strip. The required integral (the area of the strip) can then be written as

$$\int_{-h/2}^{h/2} f(x)\,\mathrm{d}x = \int_{-h/2}^{h/2} [f(0) + (x^2/2)f^{(2)}(0) + (x^4/24)f^{(4)}(0) + \ldots]\,\mathrm{d}x \tag{3.12}$$

with odd powers of x omitted from the Taylor series because they produce zero integrals. Integrating (3.12) gives the result

$$\int_{-h/2}^{h/2} f(x)\,\mathrm{d}x = hf(0) + (h^3/24)f^{(2)}(0) + (h^5/1920)f^{(4)}(0) + \ldots. \tag{3.13}$$

The term $hf(0)$ is just the midpoint estimate of the integral and the later terms represent corrections to that estimate.

To estimate the integral over N strips, between the limits L and U, we add the areas of all the strips to obtain

$$\int_L^U f(x)\,\mathrm{d}x = \sum hf(x_j) + (h^2/24)\sum hf^{(2)}(x_j) + \ldots \tag{3.14}$$

where the x_j are the midpoint coordinates of the strips and the sums are taken over all the midpoints. Instead of performing the second sum explicitly, we can recognize it to be just the midpoint rule estimate for the integral of $f^{(2)}(x)$ between L and U. This means that, with an error of order h^4, we can rewrite (3.14) as

$$\int_L^U f(x)\,\mathrm{d}x = \sum hf(x_j) + (h^2/24)f^{(1)}(x)\Big|_L^U + \mathrm{O}(h^4). \tag{3.15}$$

The higher-order terms in (3.15) involve the even powers h^4, h^6, etc. To make (3.15) resemble more closely a power series in h^2, we can formally denote by $I(h)$ the midpoint estimate obtained by using the strip width h. We can then call the exact value of the integral $I(0)$, or just I, regarding it as the result at $h = 0$. This point of view leads to a series of the form

$$I(h) = I + A_2 h^2 f^{(1)}|_L^U + A_4 h^4 f^{(3)}|_L^U + \ldots. \tag{3.16}$$

This series is applicable for the most commonly used quadrature rules, for which the quadrature points and weights are symmetrically deployed about the strip centre. The approach from first principles outlined above has been used to make the general result (3.16) plausible; it also shows that for the midpoint rule we have $A_2 = -1/24$. A_4, A_6, etc could be found by extending this simple approach, but there is a more elegant way to obtain them.

3.11 Finding the A_N

Consider the family of polynomials $F_N(x)$, of which two members are

$$F_2 = (1/2)x^2 \tag{3.17}$$

$$F_4 = (1/24)(x^4 - 2x^2). \tag{3.18}$$

The F_N are constructed to have the property (for odd M)

$$F_N^{(M)}\big|_0^1 = (M + 1 = N) \tag{3.19}$$

if we use a Boolean function form of the Kronecker delta. It follows that if we integrate F_N between 0 and 1, using the strip width h in some quadrature rule, then the difference $I(h) - I$, which is known exactly, will equal $A_N h^N$. Each coefficient A_N can thus be calculated directly by integrating F_N.

To apply the method described above we study the class of quadrature rules which estimate the area of a strip of width h by forming the product of h and the average of the f values at two points symmetrically placed about the strip centre. These rules can be applied directly to a *single* strip of width 1, and so give the single-strip estimate

$$\int_0^1 f(x)\,\mathrm{d}x = \tfrac{1}{2}[f(X) + f(1 - X)] \tag{3.20}$$

where X is used to denote the position of the first point. The single strip estimate (3.20) will, of course, be terrible for an *arbitrary* function f, but for the special polynomials F_N it will give the exact result

$$\tfrac{1}{2}[F_N(X) + F_N(1 - X)] = \int_0^1 F_N(x)\,\mathrm{d}x + A_N. \tag{3.21}$$

Using the explicit forms of F_2 and F_4 in (3.21) and rearranging gives the results

$$A_2 = (1/12)(6X^2 - 6X + 1) \tag{3.22}$$

$$A_4 = (1/720)[30X^2(1 - X)^2 - 1]. \tag{3.23}$$

The trapezoidal and midpoint rules correspond to the cases $X = 0$ and $X = \tfrac{1}{2}$ in these equations, and their A_2 and A_4 values are shown in table 3.1. If we make the special choice $X = (1/2) - (1/12)^{1/2}$ then A_2 is zero, giving a quadrature rule with no h^2 error term and a small h^4 error term. This is the traditional two-point Gaussian quadrature rule. The approach which we have adopted here allows us to check whether there is an X value for which A_4 is zero; the associated quadrature rule would have leading error terms of h^2 and h^6 type, with the h^4 term missing. It turns out that there *is* such an X value; the associated A_2 and A_4 values are given in the table. Killingbeck (1985b) gave some A_6 values and the polynomials F_6 and F_8.

3.12 A modified midpoint rule

The leading correction term for the midpoint quadrature rule can be transformed into a finite difference form by adding two extra midpoints (at $L - h/2$ and $U + h/2$) to the set of midpoints considered and using the central difference representation

$$(h^2/24)f^{(1)}(U) = (h/24)[f(U + h/2) - f(U - h/2)] \qquad (3.24)$$

with a similar representation at $x = L$ also. The result of this procedure is to produce an 'external midpoint rule' in which the two external midpoints receive the weighting factor 1/24, the two end internal midpoints receive a weight of 23/24 and the other midpoints have their usual weight of 1. The leading error is of order h^4, since the new points and weights allow for the h^2 correction explicitly. We quote the A_2 and A_4 values for this method in table 3.1, as obtained by the methods described in the preceding section.

Since the A_2 values for the trapezoidal and midpoint rules are in the ratio $-2:1$, it follows that a combination of the midpoint and trapezoidal estimates can be made which is of h^4 error type, with cancellation of the h^2 errors which arise from the two estimates. When this idea is translated into the form of a quadrature rule, the result is a weighting scheme with relative weights $1, 4, 2, 4, 2, \ldots, 4, 1$ and requiring an even number of strips. This is just the traditional Simpson's rule, which has a $|A_4|$ value roughly twice that of the modified midpoint rule described in the preceding paragraph.

Table 3.1 A_N coefficients for simple quadrature rules.

Method	X	A_2	A_4
Trapezoidal	0	1/12	−1/720
Midpoint	1/2	−1/24	7/5760
Gauss	$1/2 - (1/12)^{1/2}$	0	−1/4320
Zero A_4	$(1/2) - [(1/4) - (1/30)^{1/2}]^{1/2}$	$1/12 - (1/120)^{1/2}$	0
Simpson		0	−1/180
External midpoint		0	17/5760
9, 6 rule		0	−7/640

3.13 Relative merits of quadrature rules

It might seem obvious that quadrature rules which have relatively small coefficients A_2 and A_4 are the best ones, so that the two-point Gauss rule would be the best one in the family of rules discussed in §3.11. However, the rules with X equal to 0 or $\frac{1}{2}$ require one function evaluation per strip on average, whereas rules (such as the Gauss one) which have other X values require two function evaluations per strip, at points which have irrational coordinates. Allowing for computational effort, then, the more simple rules are not as extremely disadvantaged as at first appears.

For the case of the trapezoidal ($X = 0$) rule, the estimates using N strips and $2N$ strips have half of their sample points in common, which means that it is possible to get away with $2N$ distinct function evaluations. This point is often made in textbooks, but it is worth noting that the result of a Richardson extrapolation using the trapezoidal results for N and $2N$ strips is just the same result which would be obtained by using Simpson's rule for $2N$ strips. We can see this algebraically as follows:

$$4[f(0)/2 + f(1) + f(2) + \ldots]h - 1[f(0)/2 + f(2) + \ldots]2h \\ = h[1f(0) + 4f(1) + 2f(2) + \ldots]. \quad (3.25)$$

For the midpoint ($X = \frac{1}{2}$) rule the calculations for N strips and for $3N$ strips have function evaluations in common. Performing the algebraic Richardson extrapolation of the midpoint estimates for N and $3N$ strips leads to a rule which requires $3N$ strips and has the weights

$$9h/8 \text{ at midpoints } 1, 3, 4, 6, 7, \ldots, 3N$$

$$6h/8 \text{ at midpoints } 2, 5, 8, \ldots, 3N - 1.$$

This rule, which removes the h^2 error term of the simple midpoint rule without using any exterior midpoints, can be called the 9, 6 rule.

3.14 Three-point rules

Although the theory of §3.11 was formulated for two-point quadrature rules, it can also be used to produce some special three-point rules. For this purpose it is best to simplify the algebra by making the substitution $X = (1/2) - Y$ and by defining the quantity Z where

$$Z = X(1 - X) = (1/4) - Y^2. \quad (3.26)$$

We can then ask whether there is a value of Y not equal to zero such that the ratio A_4/A_2 equals $-7/240$, that is the value which it has for the midpoint ($Y = 0$) rule. Rewriting equations (3.22) and (3.23) for A_2 and A_4 in terms of Z leads to a simple piece of algebra which shows that the ratio A_4/A_2 has the value $-7/240$ at $Y = 0$ and at $Y = (3/20)^{1/2}$. It then follows that there must be a linear combination of the midpoint rule and the rule with $X = (1/2) - (3/20)^{1/2}$ which will give $A_2 = A_4 = 0$. Completing the calculation to find this linear combination produces the three-point integration rule which estimates the area of a strip of width h by the formula (in an obvious notation)

$$(h/18)[5f(X) + 8f(1/2) + 5f(1 - X)] \tag{3.27}$$

with $X = (1/2) - (3/20)^{1/2}$. This is the traditional three-point Gaussian rule.

If the procedure outlined above is also carried out for the trapezoidal rule, it turns out that the values $Y = (1/2)$ and $Y = (1/20)^{1/2}$ both produce a ratio of $-1/60$ for A_4/A_2. This fact leads to a quadrature rule in which the area of a strip of width h is approximated by the formula

$$(h/12)[f(0) + 5f(X) + 5f(1 - X) + f(1)] \tag{3.28}$$

with $X = (1/2) - (1/20)^{1/2}$. Both of the formulae (3.27) and (3.28) have leading error terms of order h^6, when N strips are used in sequence to estimate an integral between two limits L and U. It is important to note that this error is essentially a cumulative one. For example, the leading error per strip for the simple midpoint rule of equation (3.14) is proportional to h^3, but the leading error for the integral between L and U is proportional to h^2, since the number of strips needed to move from L to U is proportional to h^{-1}.

ROMBERG. Programming notes

3.15 The subroutine structure

The central subroutine required is one which performs one run of the integration, with the number of strips and the quadrature method both being specified by the calling module. This subroutine is always used, whereas the ones which perform the end correction for a single run or the Richardson extrapolation for two runs may or may not be used in any particular calculation.

3.16 The two-point methods

Any symmetrical two-point method can be specified by giving the appropriate X value, as explained in §3.11. Further, the h^2 correction to the estimated integral obtained from a *single* run can be computed using the theoretical results of that section, since the required A_2 coefficient is given in terms of X by equation (3.22). This means that the corrected estimate (called IC in the program) for one run is already 'boosted' to h^4 error type, even though the basic integration rule is of h^2 error type. To remove the remaining h^4 error and so end up with an

h^6-type error, it suffices to repeat the process using twice as many strips and then use Richardson extrapolation on the corrected estimates for N strips and $2N$ strips.

3.17 Three-point rules

The theory of §3.14 shows that results for symmetrical three-point quadrature formulae can be expressed as linear combinations of results for two-point formulae. In particular, the three-point Gaussian formula gives a result which is a 5 : 4 weighted average of the two-point results for $X = (1/2) - (3/20)^{1/2}$ and $X = (1/2)$. Accordingly, it is only necessary to perform the two-point integrations with the specified X values and then form the required linear combination. This can be accomplished by making appropriate calls to the subroutine concerned, with a short extra subroutine being added to form the linear combination. The result is an estimate which has h^6-type error, but which takes only two runs with N strips, instead of one run with N strips and another with $2N$ strips.

3.18 ROMBERG. Program analysis and program

Line 8 sets up a small array T which is used with the Gaussian quadrature rules and which would also be used if an external Richardson extrapolating routine were linked to the program. The variable IC (the corrected integral) is declared to be zero for later use.

Line 10 specifies the user-defined function to be integrated.

Lines 20 and 25 are input lines for the lower (L) and upper (U) limits of integration and the number N of strips to be used.

Lines 30 to 45 require 2 or 3 as input to apply the Gauss two- or three-point rules, and then produce a jump to line 80 if G = 3 is chosen.

Lines 55 to 85 specify the sequence of actions needed to work out the integral. Subroutine 200 performs a single integration. Subroutine 300 performs the end correction which removes the h^2 error and subroutine 400 performs the Richardson extrapolation. The indices J = 1 and J = 2 are used to store the successive estimates in the elements T(1) and T(2) of the T array.

Lines 90 to 120 control the application of the three-point Gauss quadrature. Two integrations with different X values are carried out; lines 95 and 110 specify the required X values. Subroutine 300, for the end correction, is not used, but subroutine 500 is used to form the correct result from the computed two-point results. Note that no change in the strip number N is involved, whereas the other choices of method involve a doubling of N (in line 70).

Lines 200 to 290 constitute the integration subroutine. Lines 200 to 215 set various parameters which depend on the values of N and X, since successive calls to the subroutine usually change at least one of these two variables. Lines 220 and 225 set up underlined headings to go above the displayed results; the printing convention used is fairly common in BASIC, but these lines might need some modification. Lines 230 to 280 form the requisite two-point sum of function values, with the multiplication by the strip width being performed last. The computed integral is stored in the T array, and printed if the Gauss three-point method is being used.

Lines 300 to 370 calculate the end correction which allows for the h^2-type error and so improves the I value obtained from a single integration. Two midpoints are used in estimating the gradient, one internal and one external, and these estimates are then inserted in the theoretical formula (line 330) explained in §3.12. Line 340 prints out both the 'raw' estimate and the corrected estimate for comparison.

Lines 400 to 430 perform the Richardson extrapolation which removes the h^4 error from the result by combining the *corrected* estimates for N and $2N$ strips.

Lines 500 to 530 are used only with the three-point Gauss rule and form the correct combination of the pair of two-point results which have been computed under the control of lines 80 to 110.

The *mathematical* form of the special X value needed in the two- and three-point Gauss rules is given in lines 40 and 95, so that the computer will work these out to whatever number of digits it uses.

```
 5 REM  ***********************************
 6 REM  ROMBERG
 7 REM  ***********************************
 8  DIM T(5): LET IC=0
 9 REM  ***********************************
10 DEF FN F(X)=EXP (-X*X)
17 REM  ***********************************
18 REM *INPUT LIMITS,METHOD,N*
19 REM  ***********************************
20 INPUT "LIMITS L,U";L,U
25 INPUT "STRIP NUMBER";N
30 INPUT "2 OR 3 PT GAUSS";G
35 IF G=3 THEN GO TO 90
40 LET XQ=1/2-SQR (1/12)
45 IF G=2 THEN GO TO 55
50 INPUT "QUADRATURE X";XQ
54 REM  ***********************************
```

```
 55 LET J=1
 60 GO SUB 200: GO SUB 300
 65 LET J=2
 70 LET N=2*N
 75 GO SUB 200: GO SUB 300
 80 GO SUB 400
 85 GO TO 20
 89 REM   ********************************
 90 LET J=1
 95 LET XQ=1/2-SQR (3/20)
100 GO SUB 200
105 LET J=2
110 LET XQ=1/2: GO SUB 200
115 GO SUB 500
120 GO TO 20
197 REM   ********************************
198 REM   INTEGRATION
199 REM   ********************************
200 LET H=(U-L)/N: LET D=H/2
205 LET YQ=1-XQ: LET AQ=(1-6*XQ*YQ)/12
210 LET X1=XQ*H+L
215 LET X2=YQ*H+L: LET S=0
220 PRINT "N";"      ";"I";"      ";"IC"
225 PRINT "--- -----          -----"
229 REM   ********************************
230    FOR M=1 TO N
240 LET A=FN F(X1)+FN F(X2)
250 LET A=A/2: LET S=S+A
260 LET X1=X1+H: LET X2=X2+H
270    NEXT M
280 LET I=S*H: LET T(J)=I
285 IF G=3 THEN PRINT N;" ";I
290 RETURN
297 REM   ********************************
298 REM   END CORRECTION
299 REM   ********************************
300 LET CU=FN F(U+D)-FN F(U-D)
310 LET CL=FN F(L+D)-FN F(L-D)
320 LET IS=IC
330 LET IC=I-AQ*H*(CU-CL)
340 PRINT N;"      ";I,IC
350 PRINT
370 RETURN
397 REM   ********************************
398 REM   RICHARDSON EXTRAP
399 REM   ********************************
400 LET IP=(16*IC-IS)/15
410 PRINT "EXTRAP";" ";IP;" , ";"X=";XQ
```

```
420 PRINT
430 RETURN
497 REM  ********************************
498 REM  3 POINT GAUSS
499 REM  ********************************
500 LET IP=(5*T(1)+4*T(2))/9
510 PRINT: PRINT "GAUSS";" ";IP
530 RETURN
531 REM  ********************************
```

ROMBERG. Specimen Results

3.19 The error integral

Table 3.2 shows some results obtained for the integral of $\exp(-x^2)$ between 0 and 1. For a quantum mechanical oscillator with the Schrödinger equation

$$-D^2\psi + x^2\psi = E\psi \tag{3.29}$$

the normalized ground-state wavefunction ψ gives the probability density $\pi^{-1/2}\exp(-x^2)$. The total probability associated with the classically allowed region $|x| \leq 1$ is thus $2\pi^{-1/2}$ times the integral which is estimated in table 3.2.

Table 3.2 Estimates of $\int_0^1 \exp(-x^2)\,dx$, using $N = 10$ and 20.

X	$I(10)$	$IC(10)$	$IP(10, 20)$
0	0.746 210 80	0.746 823 42	0.746 824 13
$(1/2)-(3/20)^{1/2}$	0.746 578 74	0.746 823 79	0.746 824 13
$(1/2)-(1/12)^{1/2}$	0.746 824 10	0.746 824 10	0.746 824 13
1/2	0.747 130 88	0.746 824 57	0.746 824 13

The numerical results illustrate several points. First, the corrected value IC is a considerably improved estimate for all four X values and the Richardson extrapolated result based on $N = 10$ and $N = 20$ is identical for them all. Second, the X value $(1/2) - (1/12)^{1/2}$ gives a two-point Gaussian quadrature; it gives a zero A_2 value (so that $IC = I$) but has a very good $I(10)$ value. Third, a little arithmetic shows that the 5 : 4 weighted average of the second and fourth $I(10)$ values is 0.746 841 3. This is, of course, the result given at $N = 10$ by the three-point Gauss rule, and this method still gives an almost exact result even at $N = 5$. This is quite remarkable; taking the appropriate linear combination of the pair of two-point results has added about five decimal digits to the accuracy.

3.20 End-point singularities

The integral of $(3/2)x^{1/2}$ between 0 and 1 is easily seen to have the exact value 1. If the integral is estimated numerically, the end-point correction subroutine has to be cut out of the sequence of operations, since it will attempt to find the square root of a negative number at the lower integration limit. The gradient $f^{(1)}$ is singular at $x = 0$, although the function f is not. This clearly spoils the simple theory presented earlier, which holds only for integrands which have continuous derivatives throughout the region of integration.

The reader may check the following facts: if the end correction at the non-singular ($x = 1$) end is included, then IC is *worse* than I as an estimate of the integral; if N is doubled the error in I is decreased by a factor of roughly 2.8 for *all* the quadrature rules, including the three-point Gauss rule.

The reason for this unexpected behaviour is that the error series for this integral with an end-point singularity has a leading term of type $h^{3/2}$. Doubling N thus decreases the error by a factor of roughly $2(2)^{1/2}$. The Richardson extrapolation subroutine can be adjusted to apply such a factor to the I values, but for some of the integrals which appear in quantum mechanics it is possible to change the variable of integration by prior analysis to produce an integral with a standard error series (i.e. $h^2, h^4, \ldots$). Killingbeck (1985a, b) discusses an example which arises in the WKB calculation of the energy levels of power law potentials. Yet another way to handle an integral which yields a non-standard error series is to fill the T array with I values obtained by repeatedly scaling up N by a fixed factor (often two) and then perform a Padé approximant analysis using the program WYNN.

We can illustrate the ideas discussed above by using an example given by Kuo (1971). He applied the three-point Gauss rule directly to the integral on the left in the equality

$$\int_0^1 (1/2)(1 - x^2)^{1/2}\,\mathrm{d}x = \int_0^1 x^2(2 - x^2)^{1/2}\,\mathrm{d}x. \tag{3.30}$$

He noted that the result for $N = 2$ was better than the result for $N = 1$, without observing that an $h^{3/2}$ error law is involved. The alternative form of the integral can be obtained by making the substitution $x = 1 - y^2$ and finally changing the dummy variable name back from y to x. In its new form the integral has a standard type of error series. Table 3.3 shows this dramatically.

Table 3.3 Three-point Gauss rule results for the two forms of the integral (3.30).

N	Form 1	Form 2
1	0.394 509 09	0.392 791 79
2	0.393 333 36	0.392 702 31
4	0.392 922 32	0.392 699 16
8	0.392 777 82	0.392 699 08
16	0.392 726 89	0.392 699 08

3.21 Residual error examples

If the integral of $\exp(-x^2)$ is carried out with $L = 0$ and $U = \infty$, then the numerical value obtained should agree with the analytical value $\pi^{1/2}/2$. Similarly, the function $(1 + x^2)^{-1}$ should give the result $\pi/2$ when integrated between 0 and ∞. Both functions have the leading terms $1 - x^2$ in their series expansion at $x = 0$. Further, every term in the formal error series for these functions vanishes, since the derivatives of odd order vanish at both limits of integration. Such integrands provide clear tests of the simple theory; they illustrate directly the existence of a residual error term which is not described by the formal error series. That series must be regarded as an asymptotic series which gives the true error with an accuracy which increases rapidly as h tends to zero. In that limit the residual error decreases more rapidly than any finite power of h.

When working to eight digits it is adequate to perform the integral of $\exp(-x^2)$ only up to $U = 5$, since the contribution from the rest of the integration out to infinity is negligible. The results are remarkable: for all the quadrature rules the correct result 0.886 226 93 is obtained for any strip number greater than 8. Nevertheless, there *is* an error for $N < 8$, and it increases rapidly as N decreases. At $N = 4$ the midpoint rule gives the estimate 0.883 025 57, while the three-point Gauss rule gives the estimate 0.886 175 57.

To integrate the slowly decreasing function $(1 + x^2)^{-1}$ out to large U values it is difficult to do the calculation by a 'head-on' direct computation. By changing variables according to the prescription $x = y \exp(y^2)$ we obtain the result

$$\int_0^\infty (1 + x^2)^{-1} \, dx = \int_0^\infty Z(1 + x^2 Z^2)^{-1}(2x^2 + 1) \, dx \qquad (3.31)$$

with $Z = \exp(x^2)$. In *numerical* computation the use of mixed expressions within an integrand is quite permissible, since the requisite quantities such as Z can be worked out numerically at each x value; in *analysis*, of course, the tradition is to express the integrand completely in terms of the new variable after a change of variable. Using the transformed integral it is only necessary to go out to $U = 5$ to get an eight-digit estimate of the infinite integral. The estimates obtained as the number of strips N is increased show an oscillating convergence to the correct result; for $N > 25$ all methods agree in giving the result 1.570 796 3.

For the case of the integral of $(1 + x^2)^{-1}$ between $L = 1$ and $U = \infty$, the substitution $x = \exp(y^2)$ is appropriate, so that the transformed limits can be taken as $L = 0$, $U = 5$. The A_2 coefficient for this case is non-zero, and for $N > 35$ the IC values given by the midpoint rule agree with the correct value 0.785 398 16. The two- and three-point Gauss rules give the correct result already at $N = 20$. The numerical experiments outlined in this section demonstrate the existence of a residual error beyond that given by the formal series in h^2, but they also illustrate that it can be eliminated by increasing N until the leading terms in that series become dominant. If the use of N strips gives a noticeable residual error while the use of $2N$ strips does not, the extrapolated estimate IP can be worse than the I value at $2N$, since the residual error varies more rapidly than the h^4 law which is used in the Richardson extrapolation. The three-point Gauss rule seems to be less subject to this kind of anomalous effect, since it uses a weighted average of results for a single N value.

3.22 Final comments

The program ROMBERG is intended to carry out the numerical integration of integrands which can be specified as mathematical functions and thus can be evaluated at the particular points used in the quadrature rule. A rule such as the trapezoidal ($X = 0$) rule will also work for integrands which are specified numerically at the points of a uniform grid, such as that used in a finite difference simulation of the Schrödinger equation. In the case of an integrand known only at some irregularly spaced set of points it is not possible to increase the number of points N and apply extrapolation or correction procedures, so the best that can be done is to use a quadrature rule which has a high-order theoretical error. Kuo (1971) gave a derivation of such a rule by means of a Taylor series approach; for the case of a uniform grid spacing the method becomes that of Simpson. Although Simpson's rule is not explicitly incorporated in ROMBERG, it corresponds to a sum of the form $(1/3)I(X = 0) + (2/3)I(X = 1/2)$. A few lines of control program can be used to apply it by making appropriate subroutine calls in the manner of those used for the three-point Gauss rule.

Phythian and Williams (1986) made the interesting observation that if the integrand is fitted by a cubic spline then integrating the spline function exactly yields a result equivalent to that obtained by applying to the original function the trapezoidal rule plus the exact A_2 correction. We should note here that the computational end-point correction method used in ROMBERG changes the value of the A_4 coefficient. The estimate of $f^{(1)}$ used includes an error of h^2 type, so that the estimate of $h^2 f^{(1)}$ has an h^4 error. This error will combine with the theoretical A_4 value to give a modified A_4 value, although this does not affect the validity of the extrapolation process which removes the h^4 error contribution.

As a further reminder of the value of theoretical analysis prior to computation, we give three examples of integrals which have been used as 'awkward cases' to motivate the development of new techniques of integration. Each integral is given first in the form originally cited and then in a converted form which leads to a normal error series of the type which ROMBERG handles with ease. The changes of variable needed are indicated below the integrals. The first two integrals involve end-point singularities.

Fox (1967):

$$\int_0^1 x^{1/2}(1-x)^{1/2}\,dx = 4\int_0^1 (2-z^2)^{1/2}(1-z^2)^2 z^2\,dz \tag{3.32}$$

$$(x = y^2,\ y = 1 - z^2,\ \text{i.e. } x = (1-z^2)^2)$$

Lyness and Ninham (1967):

$$\int_0^1 t^{-3/4}(1-t)^{-1/2} h(t)\,dt = 8\int_0^1 h(t)(2-z^4)^{-3/4}\,dz \tag{3.33}$$

$$(t = 1 - y^2,\ y = 1 - z,\ \text{i.e. } t = z^4(2-z^4)).$$

The third example involves an integrand so narrowly peaked that thousands of strips are needed to produce a good estimate of the original integral; 50 strips or less suffice after the change of variable.

Smith and Lyness (1969), $a = 0.01$:

$$\int_{-1}^1 \exp(x)(x^2+a^2)^{-1}\,dx = 3\int_{-1}^1 \exp(y^3)(y^6+a^2)^{-1}y^2\,dy \tag{3.34}$$

$$(x = y^3).$$

The references cite several works which discuss the use of extrapolation methods (in h or N) in conjunction with various types of numerical quadrature. The works of Cohen (1980) and Håvie (1966) start from the idea of Romberg integration and the trapezoidal rule. Squire (1975) gives a plentiful supply of numerical examples in a discussion of a partition–extrapolation approach. In this connection it should be noted that the idea of Romberg integration does *not* require a constant strip width h across the region of integration. Provided that the h values in *fixed* portions of the range have *fixed* ratios to one another, it is still possible to apply the extrapolation ideas by using (for example) the largest h as the relevant parameter. The reader will be able to use the simple theory of §3.10 to conclude that the A_2 coefficient will then acquire extra contributions from the $f^{(1)}$ values at each boundary across which the step length changes; the direct use of repeated integration and extrapolation without an explicit A_2 correction then becomes a more simple approach. Lyness (1972) and Evans *et al* (1983) discuss the use of extrapolation methods for the Gaussian and related quadrature formulae, and Allen *et al* (1974) establish a link between Gaussian quadrature and the theory of Padé approximants.

4 Some interpolation and extrapolation methods

Programs

INTERP, SPLINE, SPLEEN, WYNN.

4.1 General introduction

The first program of the chapter, INTERP, applies the Lagrange interpolation procedure in the Newton divided difference form, so that a simple recurrence relation algorithm can be used. The program will allow interpolation to be based on any number of consecutive points in the data set, and the data points need not be uniformly spaced. The programs SPLINE and SPLEEN use quadratic splines with uniformly spaced knots, first to perform simple interpolation and then to produce a collocation approach (devised during the writing of this book) to the calculation of the energy levels arising from the Schrödinger equation. The final program WYNN, which can be used to perform rational extrapolation of a numerical sequence or to form Padé approximants, is intended for use along with other programs. The historical survey given in §§4.19 to 4.22 brings together ideas which are scattered throughout the literature.

INTERP. Mathematical theory

4.2 The Lagrange interpolation procedure

INTERP uses the most simple type of interpolation method, which is based on fitting a polynomial of nth degree to the data given at $n+1$ points (x_j, y_j). To produce the interpolated value $y(x)$ at some point other than a data point, we can adopt either a direct or an indirect method. A direct method employs an algorithm which produces the numerical value of $y(x)$ as output when x is given as input. An indirect method produces first the explicit interpolating polynomial of nth degree and then computes $y(x)$ by substituting the x value into that polynomial. Krogh (1970) described several direct method algorithms which produce $y(x)$ accurately, but which involve repetition of some of the same computations for each chosen x value. INTERP first sets up a table of the required divided differences and then performs a short x-dependent calculation based on these stored values. The tabulation is thus only performed once.

The Lagrange interpolating polynomial of nth degree for $(n + 1)$ data points takes the form

$$L(x) = \sum y_j F_j(x) \tag{4.1}$$

where the function F_j is a polynomial of nth degree in x with the special property that it equals one at x_j and zero at all other data points. This property ensures that $L(x)$ is a polynomial of nth degree which fits correctly at the data points. $F_j(x)$ can be written in the form of a ratio of products

$$F_j(x) = P_j(x)/P_j(x_j) \tag{4.2}$$

where $P_j(x)$ is the product of n factors $(x - x_k)$ with the $(x - x_j)$ factor *omitted*. The resulting functions $F_j(x)$ clearly have the required properties.

Instead of evaluating the $F_j(x)$ and performing the sum in (4.1) it is preferable in numerical work to use compact algorithms which produce a mathematically equivalent result by using recurrence relations. Two of the classical methods of this type are those of Aitken (1932) and Neville (1934), which are described in many textbooks on numerical analysis. Krogh (1970) pointed out that Newton's divided difference formula, although widely quoted in an algebraic form, is not so often used in numerical computation, because of a popular view that it is inefficient. On a modern microcomputer, however, the calculations required to apply the formula take only a few seconds. Accordingly, INTERP has been based on Krogh's ideas, suitably modified to make use of the array storage capacity of a microcomputer. INTERP provides a good example of the use of short loop structures to perform a calculation of arbitrary length and it also involves the use of a nested multiplication process.

4.3 The divided difference approach

If the data are given in the form of the pairs (x_j, y_j) then the divided differences, which we denote by $[1, 2, \ldots, n]$, obey the following relations

$$(x_j - x_k)[j, k] = y_j - y_k \tag{4.3}$$

$$(x_j - x_l)[j, k, l] = [j, k] - [k, l] \tag{4.4}$$

and so on, in a form which can easily be adapted to use the notation of two-dimensional arrays. For example, if we set (switching to capital letters and array notation)

$$Y(J) = D(J, 1) \tag{4.5}$$

then we can use the recurrence relation

$$[X(N) - X(N + M - 1)]D(N, M) = D(N, M - 1) - D(N + 1, M - 1) \tag{4.6}$$

to place the M-term differences in the array column M. Having stored the divided differences, it is now necessary to compute the interpolated value $y(x)$ using Newton's divided difference interpolation formula, which takes the form

$$y(x) = y_1 + (x - x_1)[1, 2] + (x - x_1)(x - x_2)[1, 2, 3] + \dots. \tag{4.7}$$

That the resulting $y(x)$ is formally the same as the $y(x)$ obtained by using the Lagrange interpolation method is demonstrable by algebra, of course; here we shall concentrate on the computational use of (4.7).

The most compact way to apply (4.7) is to use a nested multiplication loop somewhat as follows; we suppose that we wish to go as far as the term [1, 2, . . ., P] and use a hybrid (but obvious) notation

```
Y:= [1,2,. . .,P]
FOR J=1 TO P-1
DX:= X-X(P-J)
Y:= Y*DX +[1,2,. . .,P-J]                    (4.8)
NEXT J
Y:= Y+Y(1)
```

The reader may check that this works by running through the algorithm for the case P = 3 of equation (4.7).

4.4 Computing derivatives

The loop structure displayed above makes possible a beautiful simplification in evaluating the *derivative* of the interpolating curve at a point x. Directly differentiating the full expression (4.7) looks complicated, but it is easy to see how the final Y value yielded by the algorithm changes if X changes. If we denote the required gradient at x by G, all we have to do is to differentiate within the assignment statement which produces the new Y value from the old one within the loop. Differentiating with respect to X leads to the extra assignment statement

$$G = G*DX + Y \tag{4.9}$$

which must come *before* the Y assignment statement to ensure that the old Y value is used in calculating the new G value. The initial value of G is zero. The second derivative GG can be computed by using the statement

$$GG:=GG*DX + 2*G \tag{4.10}$$

before the G statement, with an initial value of zero for GG.

INTERP is based on Newton's divided difference interpolation formula, and does not require the data coordinates x_j to have a regular spacing. If the x_j are set out on a grid with constant spacing h, then the divided difference method becomes identical with the method which uses forward differences. The interpolation formula for that method is

$$y(Nh + Xh) = y_N + X\Delta y_N + X(X - 1)\Delta^2 y_N/2! + \ldots \tag{4.11}$$

and is easily remembered by analogy with the Taylor series of differential calculus (to which it can be reduced if h tends to zero).

INTERP. Programming notes

4.5 Arrays

The use of the D array to hold the $[1, 2, \ldots, n]$ has been outlined in §4.3. Roughly half of the array elements are unused in the calculation, but this inefficient use of memory is tolerable because of the small size of the array required. The loop algorithm of §4.3 was given in a hybrid notation, and so the principal task is to rewrite it in a way which makes it use correctly the indices which have been used while storing the $[1, 2, \ldots, n]$ values in the D array.

4.6 Variable indices

The program has been constructed so that the operator can select indices L and U (with $U > L$). The resulting value of $y(x)$ is then that based on using only the subset of data points $X(L)$ to $X(U)$ in the interpolation calculation. This is achieved simply by selecting an appropriate triangular wedge of coefficients from the divided difference table and this capability is another of the advantages of the divided difference approach.

4.7 Derivatives

INTERP as presented here producés the estimated derivative and function value for an input x value. However, the extra lines (in the correct order) can be included to estimate higher derivatives, as explained in §4.4. The values of F and G would be needed if Newton's method of root finding were to be used in locating the zeros of the interpolating polynomial which passes through the data points.

4.8 INTERP. Program analysis and program

Lines 10 and 20 set up the X column and the D array.

Lines 30 to 70 are for the input of the (x_j, y_j) data pairs. The y values are stored in the first column of the D array (rather than in a Y column) so that the algorithm starts off correctly. The values are undisturbed by the algorithm and so can be retrieved if required.

Lines 100 to 160 form the array of divided differences as explained in the theory section. The temporary store variable T has been used to improve the appearance of the program lines; a lengthy single statement could have been used.

Lines 200 to 220 pick out the particular triangular portion of the difference table which is to be used in estimating the function at the X value chosen in line 230. The reader may check by setting out an example that the K value in line 210 is the one required to pick out the correct starting column in the D array.

Line 220 sets the initial function and gradient values for use in the nested multiplication procedure.

Lines 240 to 280 perform the nested multiplication to work out Y(X), with the name DX being introduced for the variable which holds the current coordinate difference factor.

Lines 290 to 310 print out the results using a fairly standard BASIC convention; the reader can modify these lines if necessary.

Line 320 returns to the index input line, on the supposition that varying the number of data points used or the value of X is the most likely next step.

```
 7 REM  **********************************
 8 REM  INTERP
 9 REM  **********************************
10 INPUT "NO.OF POINTS";P
20 DIM D(P,P): DIM X(P)
```

```
27 REM  *********************************
28 REM  INPUT DATA
29 REM  *********************************
30 PRINT "INPUT X,Y PAIRS"
40 FOR J=1 TO P: PRINT J
50 INPUT X,Y: PRINT X,Y
60 LET X(J)=X: LET D(J,1)=Y
70 NEXT J
97 REM  *********************************
98 REM  FORM DIFFERENCE TABLE
99 REM  *********************************
100 FOR M=2 TO P
110    FOR N=1 TO P+1-M
120 LET T=D(N,M-1)-D(N+1,M-1)
130 LET T=T/(X(N)-X(N+M-1))
140 LET D(N,M)=T
150    NEXT N
160 NEXT M
161 REM  *********************************
200 INPUT "INDICES (L<U)";L,U
210 LET K=1+U-L
220 LET Y=D(L,K): LET G=0
230 INPUT "X VALUE";X
237 REM  *********************************
238 REM  COMPUTE Y(X)
239 REM  *********************************
240    FOR J=1 TO K-1
250 LET DX=X-X(U-J)
260 LET G=G*DX+Y: LET Y=Y*DX
270 LET Y=Y+D(L,K-J)
280    NEXT J
290 PRINT "L=";" ";L,"U=";" ";U
300 PRINT "X=";X
310 PRINT "Y=";" ";Y,"G=";" ";G
320 PRINT: GO TO 200
321 REM  *********************************
```

INTERP. Specimen Results

4.9 Runge's example

Powell (1981) gave a detailed numerical study of a classic example due to Runge. The example hinges on the use of a set of $N + 1$ equidistant data points in the interval $-5 \leqslant x \leqslant 5$, with $x_1 = -5$. At these points the value of the function $(1 + x^2)^{-1}$ is given, and the value of $y(x)$ at intermediate x values is then computed by interpolation using the Lagrange polynomial (which we can accomplish using INTERP). Runge pointed out that the interpolated function $y(x)$ does *not* converge uniformly to $(1 + x^2)^{-1}$ as the number of data points is increased. Powell illustrated this by giving the interpolated value of $y(x)$ at $x = 5 - h/2$, where h is the strip width $10/N$. In using INTERP we set P equal to $N + 1$ and X equal to $10/N$. The values of Y at Powell's variable test point are given by INTERP as 1.578 720 9 at $N = 10$, −39.952 456 at $N = 20$ and 1 424.349 6 at $N = 30$. This illustrates numerically the idea that for any N there exist points at which the magnitude of the error exceeds any specified quantity.

However, we are bound to say that Runge's example also involves a remarkably silly way to carry out interpolation. It would be clear by plotting out the data values that we are dealing with a symmetric curve which rises for $x < 0$ and falls for $x > 0$. To include data from the rising portion when interpolating for $x > 0$ is clearly not sensible. If we include only data for $x > 0$ and stick to a *fixed* x value the situation is much improved, since we seek only convergence at that x value. In fact, if we go a little further and use only the last quarter of the data points (from $x = 2.5$ to 5) the use of $N = 40$ gives correctly to eight digits the function values at (for example) $x = 4.17$, 4.87 and 4.99.

To modify the program for the Runge example it is convenient to insert the temporary lines

```
22 LET H=10/(P-1)
42 LET X=-5+(J-1)*H                                  (4.12)
44 LET Y=1/(1+X*X): GO TO 60
```

Powells's variable test point then is located at 5 − H/2. On a Sinclair Spectrum microcomputer this expression can be entered directly as the X input; on other machines an assignment statement line could be written into the program. To select the last quarter of the data points at $N = 40$ (P = 41) it is sufficient to choose $L = 31$, $U = 41$.

4.10 Final comments

The program INTERP applies a traditional method which allows both the estimation of derivatives and the use of arbitrary data points. It has proved to be quite adequate for a variety of interpolation problems. INTERP actually produces an Nth degree polynomial to fit the data at $N + 1$ data points. If a polynomial of lower degree is required to approximate the data, then we are dealing with a best-fit type of problem, for which a least-squares criterion is usually employed. The program GENFIT of chapter 5 can be used for the least-squares fitting of a polynomial to a set of data values. Splines are often used for interpolation, with the cubic spline being the most popular type. The next two sections of this chapter deal with simple quadratic splines, showing how they can be used not only for interpolation but also to produce a propagator method which is applicable to the Schrödinger equation of quantum mechanics.

SPLINE and SPLEEN. Mathematical theory

4.11 Quadratic splines

The most commonly used types of splines belong to the cubic spline family, and have been used for interpolation (Greville 1970, Powell 1981, Ralston and Rabinowitz 1978), in numerical quadrature (Pythian and Williams 1986) and in the solution of two-point boundary problems (Bickley 1968, Fyfe 1969, Albasiny and Hoskins 1969, Tewarson and Zhang 1986). The use of cubic spline basis functions in solving the Schrödinger equation has been discussed by Birkhoff *et al* (1966) and Shore (1973a, b). Since a cubic spline has a continuous second derivative, it seems to be a natural choice as a basis function for second-order differential equations such as the Schrödinger equation. However, in the variational principle for the eigenvalues of the Schrödinger equation it is possible to use functions which do *not* have continuous second derivatives. The licence to do this is provided by the two equivalent forms of the integral which represents the expectation value of the kinetic energy operator. In one dimension the following result holds:

$$-\int \psi \mathrm{D}^2 \psi \, \mathrm{d}x = \int (\mathrm{D}\psi)^2 \, \mathrm{d}x \qquad (4.13)$$

if the integral is taken between the boundaries at which the wavefunction is zero. The result (4.13) allows the use of the integral on the right to define an energy expectation value for functions which have a continuous derivative $\mathrm{D}\psi$. (In fact it is actually the integral on the right which is the fundamental integral used in a formal approach which derives the Schrödinger equation as the Euler–Lagrange equation associated with a variational principle). Several authors have pointed out, by contrast with the prevailing cubic orthodoxy, that quadratic splines can be usefully applied to various problems, and can sometimes give more accurate results than those obtained using cubic splines. Khalifa and Eilbeck (1982) noted this for some two-point boundary problems, doing their numerical work on pocket computers using the BASIC language. Marsden (1974) discussed quadratic spline interpolation, comparing it favourably with cubic spline interpolation. More recently Usmani (1987) gave some simple algorithms for the construction of quadratic splines with equally spaced knots. It is on that work that the ideas used in SPLINE are based, although the algebraic formulation has been modified to simplify the computations.

4.12 The fundamental equations

We start from the assumption that the data pairs (x_N, y_N) are given and that the x_N are evenly spaced. A quadratic spline is a function which has a continuous first derivative and which is a polynomial of second degree between any two knots. If the spline is denoted by $S(x)$ then it has the property

$$S(x) = A(N)(x - x_N)^2 + B(N)(x - x_N) + C(N) \tag{4.14}$$

for $x_N \leqslant x \leqslant x_{N+1}$.

The spline function is chosen to fit the given data points (x_N, y_N), but this still leaves one parameter to be assigned in order to specify $S(x)$ uniquely. Usmani (1987) studied the case in which the gradient of $S(x)$ is given at the first data point and is set equal to the gradient of the function $y(x)$ which is being fitted. In general this seems to be an onerous requirement, but there are some interpolation problems in quantum mechanics for which this procedure is exactly appropriate, as will be explained later. For a cubic spline it is possible, in the absence of more detailed information about the derivatives of $y(x)$, to use a 'natural spine', which has zero second derivative at both of the end data points.

Since $S(x)$ is of second degree only, if obeys some simple finite difference formulae exactly. If the notation F (for function) and G (for gradient) is used to denote the values of the spline and of its derivative, then the value of G at the midpoint between two knots can be written in two different forms, one involving F values and the other involving G values. Equating these two forms gives the result

$$G(N + 1) + G(N) = (2/h)[F(N + 1) - F(N)]. \tag{4.15}$$

Here h is the uniform knot spacing and the index N refers to the knot at $x = x_1 + (N - 1)h$. The first knot coordinate is called x_1, rather than the traditional x_0, to be in accord with the usual convention for computer arrays. Equation (4.15) is the consistency relation which is the basis of our algorithm. It is also possible to make use of another equation which is derivable from (4.15), but which can alternatively be derived by equating the F and G expressions for the second derivative of F at the knot N;

$$G(N + 1) - G(N - 1) = (2/h)[F(N + 1) - 2F(N) + F(N - 1)]. \tag{4.16}$$

Usmani (1987) proposed (4.16) as being more stable computationally when a large number of knots are used, but (4.15) alone has been used in SPLINE, since it is not likely that more than 10 knots will be used in a simple interpolation process.

Since the spline function $S(x)$ is to fit the given data values at the knots, the appropriate way to use the relation (4.15) is to use the data y_N values as the $F(N)$ on the right and then compute the $G(N)$, starting from the given value of $G(1)$. The resulting $G(N)$ are the values of the gradient of $S(x)$ at the knots. From the $G(N)$ and $F(N)$ it is possible to calculate the coefficients $A(N)$, $B(N)$ and $C(N)$ appearing in the defining equation (4.14) for the spline. The relevant equations are

$$C(N) = F(N) \tag{4.17}$$

$$B(N) = G(N) \tag{4.18}$$

$$A(N) = [F(N+1) - F(N) - hG(N)]/h^2 \tag{4.19}$$

$$= [G(N+1) - G(N)]/2h. \tag{4.20}$$

Usmani (1987) gave the equation (4.19), but the equivalent form (4.20) is more simple to use and renders it hardly necessary to store the coefficients $A(N)$ explicitly. For a given x value, the appropriate N value will be obtainable using the integer part function incorporated in most versions of BASIC. The calculation of the interpolated value $S(x)$ is thus accomplished by two steps

$$N = 1 + \mathrm{INT}(x/h) \tag{4.21}$$

$$S(x) = F(N) + (x - x_N)[G(N) + (x - x_N)A(N)] \tag{4.22}$$

where $A(N)$ can be replaced if desired by the explicit expression (4.20).

4.13 The collocation method. SPLEEN

The calculation of the preceding section supposes that $G(1)$ is known, along with the $F(N)$ values. This situation does arise in some quantum mechanical calculations. An even-parity wavefunction $\psi(x)$, for example, has a known gradient of zero at $x = 0$, while the perturbed energy $E(\lambda)$ for a perturbed potential of type $x^2 + \lambda x^4$ has a known gradient at $\lambda = 0$ equal to the unperturbed expectation value $\langle x^4 \rangle$. If either quantity can be computed numerically at discrete intervals of x or λ, then the simple quadratic spline method described here can be used to obtain interpolated values at intermediate points. An example is given in the specimen results section.

When splines are used in interpolation, the values of the function to be fitted are *known* at the knots. When the *unknown* solution to a differential equation is postulated to be a spline function the $F(N)$ have to be found during the calculation. The *collocation* procedure is the one most commonly used in such a situation; it involves the imposition of the requirement that the differential equation must be satisfied at a set of collocation points. For uniformly spaced knots the collocation points are usually at the knots or at the midpoints between the knots. In the case of the quadratic spline the midpoints are used, since the second derivative is discontinuous at the knots. The usual quadratic or cubic splines used for dealing with the Schrödinger equation are the so-called B splines (Shore 1973a, Khalifa and Eilbeck 1982), but we shall illustrate the collocation method by using the simple type of quadratic spline defined in the preceding section. To make the spline function obey the Schrödinger equation at the midpoints between the knots we impose the requirement

$$\mathrm{D}^2\psi = (V - E)\psi \tag{4.23}$$

at those midpoints, with ψ taken to be a quadratic spline. The value of $\mathrm{D}^2\psi$ is constant in the region between two knots, and so its value at the midpoint $N + \frac{1}{2}$ is given by

$$\mathrm{D}^2\psi = [G(N + 1) - G(N)]/h. \tag{4.24}$$

The value of ψ at the midpoint is obtained by setting $x = (N + \frac{1}{2})h$ in (4.22) using the $A(N)$, $B(N)$ and $C(N)$ values from equations (4.17) to (4.19). The reader may confirm that after a little algebra the following recurrence relation arises from (4.23):

$$G(N + 1) = G(N) + FV(N)[(h^2/2)G(N) + hF(N)]/D(N) \tag{4.25}$$

where

$$D(N) = 1 - (h^2/8)FV(N) \tag{4.26}$$

and

$$FV(N) = V(N + \tfrac{1}{2}) - E \tag{4.27}$$

in an obvious notation. From the relation (4.15) it is clear that there is also a recurrence relation for the F values,

$$F(N + 1) = (h/2)[G(N + 1) + G(N)] + F(N). \tag{4.28}$$

The equations (4.25) and (4.28) allow us to propagate the values of $F(N)$ and $G(N)$ along the x axis, if we are given E, $F(1)$ and $G(1)$. The test $F = 0$ or $G = 0$ can then be applied at the outer boundary to locate the E values which are approximate eigenvalues of the Schrödinger equation (4.23). This propagator-shooting calculation fits in with the root-finding approach, and a short subroutine SPLEEN (spline energy) can be written to serve as the appropriate function subroutine to be used in conjunction with ROOTSCAN and SECANT. The calculation can also be represented as a generalized matrix eigenvalue problem if the equations are suitably rearranged, but for the purpose of estimating the eigenvalues the propagator approach used in SPLEEN is much easier to apply. It also has the advantage of allowing the use of boundary conditions which involve arbitrary combinations of the wavefunction and its gradient, although it is not as accurate as the finite difference algorithm FIDIF of chapter 8.

Since both the programs SPLINE and SPLEEN are short and simple in structure, there is no need for a detailed programming notes section; comments about programming details are made when necessary within the description of the listed programs.

4.14 SPLINE. Program analysis and program

Lines 10 to 40 set up F and G arrays and compute the knot spacing H from the upper (U) and lower (L) values of x and the number of data points ND.

Lines 110 to 140 are the input lines for the F(N) values.

Line 150 is the input line for G(1).

Lines 160 to 190 compute the G(N), using equation (4.15) of the text. The difference quantity DIF is not essential, but has been used to improve the appearance of the printed program.

Lines 200 to 280 carry out the interpolation when the coordinate X is given. NH is the number of H units included in X and XN is the residual coordinate called $(x - x_N)$ in the text. Line 215 allows for the special case X = U. Line 230 works out the coefficient A(NH) of the local spline function, using equation (4.20). Line 240 works out the value of the quadratic spline function S(X), while line 250 works out the derivative DS(X) of the spline function. Lines 260 and 270 give neat displays of the results.

Line 205 (remmed) represents one possible way of ending the calculation by using the special input 999. In Sinclair BASIC an input of an undefined variable symbol such as 'Q' suffices to break off the calculation.

```
  7 REM  **********************************
  8 REM  SPLINE
  9 REM  **********************************
 10 INPUT "X LIMITS,L,U";L,U
 20 INPUT "NO.OF POINTS";ND
 30 DIM F(ND): DIM G(ND)
 40 LET H=(U-L)/(ND-1)
 41 REM  **********************************
100 PRINT "INPUT F VALUES"
```

```
101 REM  ***********************************
110 FOR J=1 TO ND
120 PRINT J: INPUT X
130 PRINT X: LET F(J)=X
140 NEXT J
141 REM  ***********************************
150 INPUT "G(1)";G(1)
151 REM  ***********************************
152 REM  COMPUTE G VALUES
153 REM  ***********************************
160 FOR N=1 TO ND-1
170 LET DIF=F (N+1)-F(N)
180 LET G(N+1)=(2/H)*DIF-G(N)
190 NEXT N
191 REM  ***********************************
192 REM  INTERPOLATE
193 REM  ***********************************
200 INPUT "X VALUE";X
205 REM IF X=999 THEN GO TO 300
210 LET NH=1+INT ((X-L)/H)
215 IF X=U THEN LET NH=ND-1
220 LET XN=X-NH*H+H
236 LET A=(G(NH+1)-G(NH))/(2*H)
240 LET S=F (NH)+XN*(G(NH)+XN*A)
250 LET DS=2*A*XN+G(NH)
260 PRINT "X=";X
270 PRINT "S=";S,"DS=";DS
280 PRINT: GO TO 200
281 REM  ***********************************
290 STOP
291 REM  ***********************************
```

SPLINE. Specimen results

4.15 Energy level interpolation

The ground-state energy for the perturbed oscillator Hamiltonian $-D^2 + x^2 + \lambda x^4$ can be calculated numerically as a function $E(\lambda)$ of λ using several of the programs of this book. Table 4.1 shows the input data used in an interpolation calculation in which both SPLINE and INTERP were used. Only six significant digits were retained throughout.

Table 4.2 shows some results obtained using SPLINE and INTERP with the six data points of table 4.1. The fifth-degree polynomial function of INTERP performs better than the simple quadratic spline when the theoretically correct $G(1)$ value of 0.75 is used in SPLINE. However, when the parameter $G(1)$ is treated as a variable parameter and set at

0.70, the results from SPLINE become of similar accuracy to those from INTERP. If the value of $E(0.5)$ is known and used as a reference value to set $G(1)$, then the entire set of interpolated values is much improved.

Table 4.1 $E(\lambda)$ values for the interpolation calculation.

λ	E
0.0	1.000 00
0.2	1.118 29
0.4	1.204 81
0.6	1.275 98
0.8	1.337 55
1.0	1.392 35

Table 4.2 Interpolated values for $E(\lambda)$.

λ	INTERP	$G(1) = 0.75$	$G(1) = 0.70$	Exact
0.1	1.064 78	1.067 07	1.064 57	1.065 29
0.3	1.164 15	1.161 57	1.164 07	1.164 05
0.5	1.241 80	1.244 22	1.241 72	1.241 85
0.7	1.307 81	1.305 35	1.307 85	1.307 75
0.9	1.365 51	1.368 05	1.365 55	1.365 67

4.16 SPLEEN. Program analysis and program

The program is given in a short listing form, with ROOTSCAN and SECANT regarded as already-existing modules which can be included at will in the centre of the program.

Line 10 defines the potential function, taken to be $x^2 + x^4$ in this specimen case.

Lines 20 to 40 calculate the knot separation H and precompute two quantities which are used repeatedly in the subsequent computation.

Line 50 allows the operator to specify the particular boundary conditions to be used at the upper boundary U.

Line 60 sets a scaling factor SF, typically 1 or $\frac{1}{2}$ or $\frac{1}{4}$, which is used to ensure that the F and G values do not cause computer overflow when NS is large.

Line 100 sets the initial conditions. For an even-parity potential with the lower limit L equal to zero, the choice (1, 0) and (0, 1) for (F, G) give even- and odd-parity wavefunctions, respectively.

Lines 1000 to 1100 constitute the function subroutine.

Line 1000 resets the starting F and G values before each traverse of the x axis, since the algorithm is set up to use only the scalar variables F and G, with no arrays.

Line 1020 sets the X values to be used in the collocation process as the midpoints between the knots. The reader may check that lines 1030 to 1070 apply the algorithm of the text (§4.13). The F(N + 1) calculation in line 1060 *must* precede the G(N + 1) calculation in line 1070, to ensure that the same current G(N) value is used on the right-hand side of the assignment statements.

Line 1075 applies the scaling factor to both F and G, since they both appear linearly in the calculation of the next pair of values (F, G).

Line 1090 forms the appropriate combination of the F and G values at X = U to give the function value which is returned by the subroutine. The same name F is used for the returned function and for one of the running variables in the loop. This is slightly sinful, but was originally intended to be appropriate for the case of Dirichlet boundary conditions, where the final F value is indeed the required function.

```
   7 REM  **********************************
   8 REM  SPLEEN
   9 REM  **********************************
  10 DEF FN V(X)=X*X*(1+X*X)
  11 REM  **********************************
  20 INPUT "LIMITS,L,U";L,U
  30 INPUT "NO.OF STRIPS";NS
  40 LET H=(U-L)/NS: LET H2=H*H/2: LET H8=H*H/8
  50 PRINT "BOUNDARY CONDITION": PRINT "0=DIR*F+(1-DIR)*G": INPUT
"DIR";DIR: CLS
  60 INPUT "SCALING FACTOR";SF
  99 REM  **********************************
 100 INPUT "INITIAL F,G";F0,G0
 110 INPUT "E0,DE";E0,DE
 297 REM  **********************************
 298 REM  ROOTSCAN
 397 REM  **********************************
 398 REM  SECANT
 487 REM  **********************************
 488 REM  RETURN TO ROOTSCAN
 491 REM  **********************************
 998 REM  PROPAGATOR
 999 REM  **********************************
1000 LET F=F0: LET G=G0
1010    FOR N=1 TO NS
1020 LET X=(N-1/2)*H+L
```

```
1030 LET FV=FN V(X)-E
1040 LET FG=FV*(H2*G+H*F)
1050 LET FG=FG/(1-H8*FV)
1060 LET F=F+H*(G+G+FG)/2
1070 LET G=G+FG
1075 LET F=F*SF: LET G=G*SF
1080   NEXT N
1090 LET F=DIR*F+(1-DIR)*G
1091 REM  *********************************
1100 RETURN
```

SPLEEN. Specimen results

4.17 Energy level calculations

For the Hamiltonian $-D^2 + x^2 + x^4$ the energy levels can be calculated by several different methods, and so accurate values are available to check the characteristics of the collocation-shooting program SPLEEN. Results obtained for the first two levels of even and odd parity (with L = 0, U = 5 and NS = 50, 100 and 200) gave accurate energies when the standard Richardson extrapolation based on h^2 and h^4 error terms was used. For the case L = 0, U = 2, however, it is to be expected that the confining effect of the boundary at $x = 2$ will tend to change the ground-state energy from the value of 1.392 351 6 which it has for the case U = 5. Table 4.3 shows the results for the case of Dirichlet (F = 0) and Neumann (G = 0) boundary conditions at $x = 2$, for the lowest state of even parity.

Table 4.3 Results for L = 0, U = 2. (Common digits in parentheses at top of columns.)

NS	F = 0	G = 0
	(1.397)	(1.385)
50	9542	5734
100	8636	4757
200	8410	4513
(50,100)	8334	4431
(100,200)	8335	4432

The results of table 4.3 show the different effects which the imposition of Dirichlet and Neumann boundary conditions have on the energy. The error law is an almost perfect h^2 one, as is indicated by the Richardson extrapolated results given as the last two numbers in each column. Thus, although the collocation–propagator method based on the simple quadratic spline involves a very simple algorithm, it can give accurate energy levels when used in conjunction with Richardson extrapolation of the standard type.

To achieve some degree of variety the case of quadratic (rather than cubic) splines has been treated in this chapter. However, it should be noted that for the case of equally spaced knots the theory of interpolation based on natural cubic splines can be given a fairly simple form. Greville (1970) gave an account of the theory together with some numerical coefficients which can be used to construct the interpolating splines. Grant and Burke (1967) used quadratic splines in their work on atmospheric reflection, and gave the relevant formulae for the case in which the knots are not uniformly spaced.

WYNN. Mathematical theory

4.18 The Aitken transformation

The roots of the quadratic equation

$$x^2 - 3x + 1 = 0 \tag{4.29}$$

also obey the equation

$$x = (1 + x^2)/3. \tag{4.30}$$

This mathematical equation is only satisfied by *two* particular numbers. However, if the equality sign = is replaced by the assignment statement symbol := then (4.30) becomes a rule for an iterative process in which *any* input x value on the right will produce an output x value on the right. A simple loop program with some initial value x_0 for the variable x will produce a sequence of x values; if our aim is to solve the equation (4.29) then we must arrange for the sequence of x values to converge to a root of the equation. This means that we wish to continue repeating the loop until output equals input (to within machine accuracy). A little calculus shows that if the input x value equals $r + \varepsilon$, where r is a root of equation (4.29) and ε is an error (which we take to be small, in the sense that $\varepsilon^2 \ll |\varepsilon|$) then the output x value is $r + 2r\varepsilon/3$. This means that for small ε we have a *first-order* iterative process. The output error is directly proportional to the input error, with the sequence of errors forming a geometric progression.

A short loop program for the example above shows that the input $x_0 = 0$ generates a sequence of x values converging to the root 0.381 966 01. As $|x_0|$ is increased convergence to the same root occurs until $|x_0|$ equals the other root (2.618 033 0). For greater $|x|$ values the process does not converge but gives a rapidly increasing sequence.

One way to cure this problem is to introduce a *relaxation parameter* P and use the assignment statements

```
Y:=(1+X*X)/3                                                    (4.31)
X:=P*X+(1-P)*Y                                                  (4.32)
```

With $P = 2$, the iterative process now converges to the larger root for a range of x_0 values which extends roughly from 0.4 to 4.6, with divergence outside that range. The original calculation, giving the lower root, has $P = 0$.

The idea of the Aitken transformation is to use a fixed formula which will always reduce a first-order iterative process to a *second-order* one, in which the output error is a constant times the square of the input error. The formula is fixed in the sense that no *ad hoc* adjustable parameters such as the relaxation parameter P are involved. For three successive members of the sequence of x values arising from a first-order iterative process we have

$$x_1 = r + \varepsilon \tag{4.33}$$

$$x_2 = r + k\varepsilon \tag{4.34}$$

$$x_3 = r + k^2\varepsilon \tag{4.35}$$

for some common ratio k which is usually unknown. It is quickly shown that the following two equations hold:

$$(x_3 - x_2) = k(x_2 - x_1) \tag{4.36}$$

$$(kx_1 - x_2) = r(k - 1). \tag{4.37}$$

As a computational procedure it would suffice to use the variables R and K in a program and use these last two equations to produce a *numerical* value for R. If the equations are solved explicitly by algebra we obtain the Aitken extrapolation formula

$$r = (x_1x_3 - x_2^2)/(x_1 + x_3 - 2x_2). \tag{4.38}$$

This result is easily remembered, but a more stable form of it is

$$r = x_1 - (x_1 - x_2)^2/(x_1 + x_3 - 2x_2). \tag{4.39}$$

The second form of the expression is less prone to round-off errors when x_1, x_2 and x_3 are very close together in value.

Most of the sequences produced by an iterative process do not give a perfect geometric progression of errors; if they did one application of (4.39) would give the required limit directly. In most applications a *sequence* $\{r_n\}$ of r values is produced, each r value being obtained by applying (4.39) to the preceding three elements of the $\{x_n\}$ sequence. In many cases the sequence $\{r_n\}$ converges markedly more quickly than

$\{x_n\}$ to the desired limit. Indeed, for an exact geometric sequence (4.39) gives the correct limit even when $k > 1$, that is when the sequence $\{x_n\}$ diverges. This remarkable property of sometimes producing a converged result from a divergent initial sequence is shared by the more general transformations carried out by the program WYNN; it is useful in dealing with some of the divergent perturbation series which are encountered in quantum mechanics. The program HYPOSC can be used to provide examples of such series.

If the sequence $\{x_n\}$ of values produced by the repetition of the assignment statement

```
X:=(1+X*X)/3                                                    (4.40)
```

is to be treated by the Aitken transformation we must retain the last three values. This can be arranged, for example, by using an array X(N) together with the statements

```
X:=X(N)                                                         (4.41)
X(N+1):=(1+X*X)/3                                               (4.42)
```

so that X(1) is the initial value. This is usually denoted by x_0 in mathematical treatments, but some dialects of BASIC do not allow zero as an array index. It should now be clear how to express the element R(N) in terms of the preceding three elements X(N), X(N – 1) and X(N – 2) using the equation (4.39), taking care that the calculation only starts after X(3) has been computed. Applying this procedure to the X(N) arising from (4.41) and (4.42) leads to a sequence R(N) which converges rapidly to the lowest root of the quadratic equation if X(1) has an absolute value less than that of the larger root. The convergence of the R(N) is more rapid than that of the X(N). Although the fact that the larger root cannot be calculated is disappointing, an interesting effect can be observed if the X(1) value is close to that root. For example, if X(1) is 2.62 the R(N) values stay close to 2.618 for several iterations, even though the X(N) sequence is clearly moving away from the greater root.

4.19 Multi-component sequences

The traditional Aitken transformation serves to handle sequences with errors which form a single geometric progression. However, several workers, most notably Shanks (1955), have looked for the appropriate formulae to calculate r when the sequence involves errors which involve two or more geometric progressions, so that we have the law

$$x_n = r + a_1 k_1^n + a_2 k_2^n + \ldots . \tag{4.43}$$

When only a_1 is non-zero we know that three x_n are needed (with three unknowns involved). When a_1 and a_2 are non-zero, five successive x_n are needed to calculate r (with five unknowns involved). For the case of an arbitrary number of components Shanks (1955) showed that the algebraic solution of the relevant simultaneous equations leads to an expression for r which is the ratio of two determinants, and used this expression in his numerical examples. Wynn (1956, 1966) later showed how two simple repetitive algorithms would produce results mathematically equal to the ratio of determinants used by Shanks. Wynn's algorithms lead to simple loop programs for computer use. In their most compact form they produce from the numerical sequence $\{x_n\}$ a family of estimates of r based on the assumption that the x_n contain $1, 2, 3, \ldots, N$ terms of the type in equation (4.43), where $N + 3$ is the number of successive terms of the sequence $\{x_n\}$ which are available for analysis. Clearly the hope behind such an analysis is that the estimates of r will stabilize to some well defined value as the number of components allowed for in the analysis is increased.

One simple way to attack the problem of a two-component error might seem 'obvious': we could just apply the Aitken transformation *twice*, intending to remove one component at a time. Shanks (1955) showed that this strategy does not succeed in removing both components completely; only the more complicated e_m transformation which he derived will do that. Given Wynn's algorithms, however, the e_m transformation of Shanks is nowadays easy to apply; the program WYNN includes an adjustable parameter so that it can apply a whole family of transformations (Vanden Broeck and Schwartz 1979, Hamer and Barber 1981, Killingbeck 1988a) of which the repeated Aitken transformation and the Shanks e_m transformation are special cases.

4.20 Padé approximants

The e_m transformation of Shanks acts on a sequence of numbers, which may be produced by a variety of mathematical processes. In the study of the convergence properties of a power series the principal quantity of interest is the sequence of partial sums, with x_n being the sum of the first n terms of the series. Given the coefficients of a power series we can attempt to fit the series to a rational fraction as follows

$$\begin{aligned}\sum A_m \lambda^m &= P(M)/Q(N) \\ &= \sum_0^M P_n \lambda^n \Big/ \sum_0^N Q_n \lambda^n. \end{aligned} \tag{4.44}$$

If the convention $Q_0 = 1$ is adopted, then (barring exceptional cases) it is possible to find a unique set of coefficients P_n and Q_n in (4.44) such that the expansion of the ratio P/Q agrees with the fitted power series up to the λ^{N+M} term. The resulting rational fraction is called a *Padé approximant* of the original series. We shall denote the Padé approximant in (4.44) by (M/N) (with/for division), although the conventional notation originated by Padé uses the reversed form [N|M]. When the fitted series represents a function of λ the $(M/0)$ approximants are simply the partial sums of the Maclaurin series expansion of the function.

The link between the mathematical ideas outlined in the discussion so far is as follows. For a fixed λ the Padé approximant can be evaluated numerically to give a numerical estimate of the 'sum' of the original series (even for cases in which the series diverges). The result obtained is the *same* as that obtained by applying the appropriate Shanks e_m transformation to the numerical sequence of partial sums of the series (for the same λ value, of course). This means that Wynn's algorithms can be used to produce a numerical estimate for the sum of a series *without* explicitly finding the coefficients in the polynomials P and Q.

4.21 Series of Stieltjes

A large class of power series (series of Stieltjes) can be represented formally in terms of an integral

$$I(z) = \int_0^\infty (1 + zt)^{-1}\, \mathrm{d}\phi(t) \tag{4.45}$$

where $\phi(t)$ is an appropriately defined positive measure function. We can obtain sufficient generality for our purposes by visualizing $I(z)$ as an expectation value for some quantum mechanical wavefunction. If we expand $(1 + zt)^{-1}$ as a series and formally integrate term by term, we obtain a result of the form

$$I(z) = \int_0^\infty (1 + zt)^{-1} \psi^2(t)\, \mathrm{d}t = \sum_0^\infty (-1)^n \mu_n z^n \tag{4.46}$$

where

$$\mu_n = \int_0^\infty t^n \psi^2(t)\, \mathrm{d}t = \langle \psi | t^n | \psi \rangle \tag{4.47}$$

is the nth moment (or the expectation value of t^n). In the case

$\psi = \exp(-t/2)$ the coefficient μ_n will be n! The series in (4.46), the Euler series, then diverges rapidly, even though it still represents in some sense the function of z given by the integral. Applying Padé approximant methods to the series can produce a numerical 'sum' for the series which equals the numerical value of the integral.

If the function $(1 + zt)^{-1}\psi^2(t)$ is denoted by $Y(t)$ then a little algebra shows that for any function $f(t)$ the quantity

$$A(f) = 2\langle f|\psi\rangle - \langle f|(1 + zt)|f\rangle \tag{4.48}$$

(written in Dirac notation) has the following property:

$$\begin{aligned} A(Y + \varepsilon) &= \langle \psi|(1 + zt)^{-1}|\psi\rangle - \langle \varepsilon|(1 + zt)|\varepsilon\rangle \\ &= I(z) - \langle \varepsilon|(1 + zt)|\varepsilon\rangle. \end{aligned} \tag{4.49}$$

This shows that for $z > 0$ the use of any trial function $f(t)$ in $A(f)$ produces a lower bound to $I(z)$. If $f(t)$ takes the special form of a product of $\psi(t)$ and a polynomial $F(zt)$ in powers of zt, the expression for $A(f)$ becomes

$$2\langle \psi|F|\psi\rangle - \langle \psi|(1 + zt)F^2|\psi\rangle. \tag{4.50}$$

The specially valuable feature of this expression is that it can be evaluated *without knowing* $\psi(t)$, since it only requires knowledge of the μ_n, that is of the coefficients in the power series. This leads to a procedure which only involves operations on the series and so is decoupled from any explicit knowledge of the integrand which generates the series. If the polynomial F is varied to produce a maximum of A using a fixed even number of the coefficients μ_n, then the *best* lower bounds to $I(z)$ obtained are the same as those obtained by fitting the appropriate $(N - 1/N)$ Padé approximant to that particular number of coefficients. For example, using values up to μ_5 gives a best lower bound equal to the value of (2/3). An extension of the variational approach used here (Killingbeck 1985a) leads to the conclusion that the best *upper* bounds using a fixed odd number of terms are given by the values of the corresponding (N/N) approximant. The conditions under which the (N/N) and $(N - 1/N)$ approximants 'meet in the middle' to produce a precise result as N increases have been discussed by Baker (1965) and in various later works by Baker and others. Common (1968) discussed the situation when z is negative. Longman (1971) gave a recursive algorithm which will produce the coefficients of the P and Q polynomials from the μ_n, and Levin (1973) gave some sequence transformation methods which generalize and improve upon those of Shanks (1955). Levin's main argument is that the Shanks model of a sequence as involving a set of geometric progressions, in accord with equation

(4.43), is not sufficiently general to describe some types of sequence which arise in applied mathematics. We may note here that the series of Stieltjes (4.46) represents a variant of the Shanks model in that an infinite number of geometric sequence components is involved in each term of the *series*.

4.22 The algorithm

The algorithm of Wynn which is commonly used is easily explained in terms of a staggered array

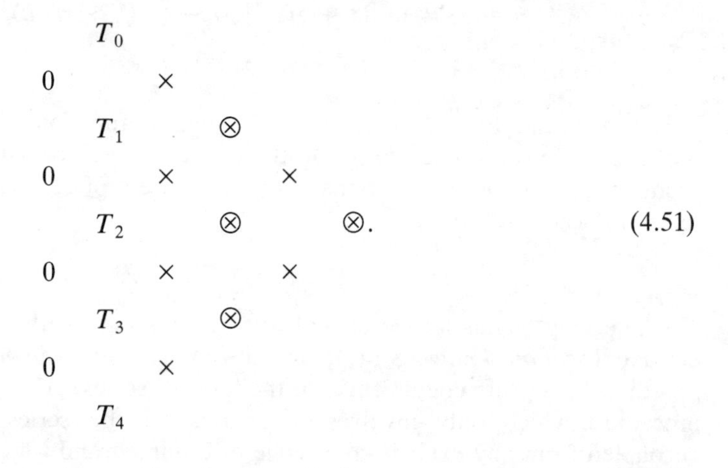

In the array the T_n are the elements in the sequence being studied. T_n might be the partial sum of a power series up to the λ^n term, or it might be the nth value resulting from some iterative process. The first two columns consist of zeros and of the T_n values. All subsequent columns are computed by using the 'lozenge algorithm'

$$\begin{matrix} & B & \\ A & & \\ & C & \end{matrix} \qquad D = A + (C - B)^{-1}. \tag{4.52}$$

The points marked × are dummy values needed at intermediate stages, while the points marked ⊗ are the desired approximant values. If the T_n are the sums up to λ^n of a power series, then the four approximants shown in the diagram are (1/1), (2/1), (3/1) and (2/2). The type of each approximant is easily decided by checking how many T_n values have contributed to its value.

The particular algorithm used in the program WYNN has several modifications incorporated in it. When a × column has been calculated it is multiplied by a factor KS (K summation) before the calculation continues. If KS is one the result is the usual Padé approximant sequence. If KS is zero then the result is the repeated Aitken transformation sequence. This arises because applying the Aitken transformation to the T_n sequence provides exactly the $(N/1)$ Padé approximants. Replacing the $\times$ elements by zero thus keeps on applying the Aitken transformation by putting us back in the situation which held at the extreme left of the table. However, the parameter KS can be set at other values in an empirical manner. The choice KS = -1 has been found to be effective in dealing with some series from quantum mechanical perturbation theory (Killingbeck 1988a). The case KS = 0 was discussed by Shanks (1955), who pointed out that the sequence of (N/N)-type elements obtained often converges quickly as N increases, even when the sequences $(N/1)$, $(N/2)$, etc, do not.

The algorithm as displayed in the diagram represents a calculation of the approximants (M/N) with $M > N$. By adding an extra zero at the top of the T_n column it is possible if desired to add a further set of points above the diagonal. This extra layer of points gives the $(N-1/N)$ approximants. If the T_n are the partial sums of a series of Stieltjes the values of $(N-1/N)$ and (N/N) straddle the value of the integral $I(z)$, at least for KS = 1. Some of the series arising in quantum mechanical perturbation theory are known to be of Stieltjes type, although the application of the Padé approximant method has been found to be effective for many sequences and series which are not of this type. It is possible for a series to be of Stieltjes type without the upper and lower bound approximant values meeting to produce a common limit; the energy perturbation series for the perturbed potential $x^2 + \lambda x^8$ has this difficult property.

WYNN. Programming notes

4.23 Setting array dimensions

The lozenge algorithm is such that to calculate column $N+2$ in the table only the values of the numbers in the preceding two columns N and $N+1$ are needed. Further, it is possible to overwrite column N with the newly calculated column $N+2$ without disrupting the continuation of the calculation, although preceding approximants will be lost. This means that an algorithm based on two linear arrays rather than on a single square array can be devised. If the starting column of $T(N)$ elements required in the algorithm is to be read in as data by the program, then the operator could quote the number of elements directly. However, WYNN, as listed here, looks for the end of the list of

$T(N)$ values automatically *but* does so on the assumption that there are some zero (i.e. unused) elements at the top of the T array.

4.24 Use as a library subroutine

WYNN may either be incorporated explicitly in other programs or be recorded separately and then combined with other programs in the microcomputer by using a MERGE instruction (on a Sinclair Spectrum) or a SPOOL instruction (on a BBCB). The latter approach means that WYNN can be used to perform a Padé analysis of the output from any program which generates numerical sequences and stores them in an array named T. The line numbers of WYNN have been made high, so that they will not interfere with those in any preceding program of reasonable length. Alternatively a short input routine may be prefixed to WYNN so that the values of the elements $T(N)$ can be typed in; another approach is to type in the coefficients $E(N)$ of a series and then use a variable λ input so that the partial sums $T(N)$ of the series are formed for the particular λ value selected. However, the Padé analysis performed by WYNN can be applied to any sequence of numbers, whether or not they represent the partial sums of a power sequence. For example, if the strip widths h, kh, k^2h, ... (with k typically 1/2) are used in some integration rule to estimate a definite integral, then the resulting estimates of the integral form a numerical sequence which can be analysed using WYNN in order to obtain an estimate of the limit of the sequence (i.e. the numerical value of the integral). This approach to numerical integration has been discussed by Kahaner (1972) and compared with the Romberg method (see the program ROMBERG).

4.25 Labelling the approximants

What WYNN produces are the numerical values of various (M/N) Padé approximants to the sum of the series for which the partial sums are given as the input $T(N)$ array. Even when the $T(N)$ are *not* the partial sums of a series we can still label the resulting (M/N) by the same rules which would apply for that case. As WYNN uses the algorithm of §4.22 to work out the (M/N) values it assigns the correct M and N labels to each number which is computed and displayed. Note that the initial partial sums are themselves Padé approximants with the second index zero; the denominator polynomial Q in that case is the number 1, as follows from the definition of Padé approximants given in §4.20. The basic principle behind the labelling is the same as that used to explain the example given in §4.22.

The labelling scheme is as follows:

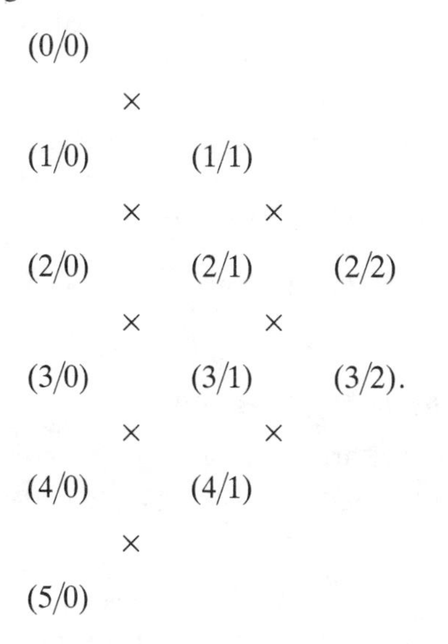

(4.53)

4.26 Storing and displaying the (M/N)

The (M/N) could be stored in a square array, but the version of WYNN listed here prints them out on the screen without saving them permanently. This is because WYNN will often be used in conjunction with programs such as HYPOSC which themselves use large arrays; the avoidance of large arrays in WYNN then makes it less likely that the microcomputer's memory capacity will be exceeded by the requirements of the composite program.

4.27 Rounding errors and scaling tests

The algorithm involves many stages of subtraction and division, as shown by equation (4.52), and the computed numbers tend to alternate between large and small values on passing from one column to the next in the lozenge algorithm. This means that rounding errors might be important for the higher-order approximants. A study of the calculation quickly shows that if the arithmetic used is exact then the only effect of multiplying all the input numbers $T(N)$ in the first non-zero column by the constant scaling factor SF is to multiply all the output values (M/N) by the same factor SF. If the (M/N) values are then *divided* by SF before being displayed or stored the results will be *independent* of the value of SF. However, in the presence of rounding errors the results *do* vary slightly with SF; WYNN allows SF to be varied in order to obtain a rough empirical estimate of the number of decimal digits which are reliable in a given (M/N) value. Experience indicates that it is only for the case of strongly divergent initial sequences, with terms varying over many orders of magnitude, that rounding errors become appreciable. Their effect can be observed, for example, in an analysis of the energy series produced by the program HYPOSC when the renormalization parameter K is set equal to zero; the resulting Rayleigh–Schrödinger series is strongly divergent (see the specimen results section).

4.28 WYNN. Program analysis and program

Lines 2000 and 2010 set up the arrays A and B to be large enough to go as far as the highest-order approximant possible. Fixed dimensions could be used.

Lines 2020 to 2060 copy the sequence to be analysed (which must be called T in the feeding program) into the B array and set the A array to zero. A and B are thus the first two rows in the algorithmic scheme explained in §4.22. Line 2020 sets the extra top element in the B column to zero in order to produce the (N – 1/N) approximants; if this leads to an early division by zero then IE−8 can be used instead. Line 2040 checks for the end of the list of T(N) elements, since the T array might not have as many elements as are allowed for by the dimension of A and B. (If the dialect of BASIC used will allow a jump out of a loop then we can use a command such as GO TO 2070, with line 2075 deleted, carrying forward the current N value in the loop).

Line 2070 sets the scaling factor SF at 1, a value which can be adjusted if required. I is a counting index used in constructing the table of approximant values.

Line 2080 sets the KS parameter, with a reminder of the two special values which give standard transformations.

Lines 2090 to 2180 calculate the elements of the approximant table, but use only the columns A and B in a cyclic fashion, since the values (together with their labels) are printed out on the screen by line 2160. Line 2140 applies the KS value to the A array to produce the desired type of transformation. Line 2150 is a specimen line which picks out the (N/N) approximants for printing; these approximants often give the best convergence. To turn off this line a REM can be attached at the start.

Some special ‘difficult’ cases exist of sequences which might cause difficulty with divisions by zero. This problem can often be overcome by separating out the calculation of the denominator, as illustrated by lines

2095 and 2125.

```
1000 DIM T(50)
1010 LET T(1)=1: LET SG=1
1020 FOR J=1 TO 20
1025 LET SG=-SG
1030 LET T(J+1)=T(J)+SG/(J+1)
1040 NEXT J
1997 REM  *********************************
1998 REM  WYNN
1999 REM  *********************************
2000 INPUT "NO.OF TERMS";Q
2010 DIM A(Q+1): DIM B(Q+1)
2017 REM  *********************************
2018 REM  COPY T INTO B
2019 REM  *********************************
2020 LET B(1)=0
2030 FOR N=1 TO Q
2040 IF T(N)=0 AND T(N+1)=0 THEN GO TO 2060
2050 LET A(N)=0: LET B(N+1)=T(N): LET NT=N
2060 NEXT N
2064 REM  *********************************
2070 LET SF=1E0: LET I=1
2075 LET N=NT
2079 REM  *********************************
2080 INPUT "1 PADE,0 AITKEN";KS
2087 REM  *********************************
2088 REM  DUMMY COLUMN
2089 REM  *********************************
2090 FOR M=1 TO N-2
2095 LET D=B(M+1)-B(M): IF D=0 THEN LET D=1E-8
2100 LET A(M)=A(M+1)+1/D: NEXT M
2101 REM  *********************************
2110 LET N=N-1: IF N=2 THEN GO TO 2190
2117 REM  *********************************
2118 REM  PADE COLUMN
2119 REM  *********************************
2120 PRINT "": FOR M=1 TO N-2
2125 LET D=A(M+1)-A(M): IF D=0 THEN LET D=1E-8
2130 LET B(M)=B(M+1)+1/D
2140 LET A(M)=A(M)*KS
2150 IF M<>2 THEN GO TO 2170
2160 PRINT B(M)/SF, "(";M+I-2;"/";I;")"
2170 NEXT M
2171 REM  *********************************
2180 LET N=N-1: LET I=I+1: IF N>2 THEN GO TO 2090
2190 PRINT "GOTO 2020 TO REPEAT"
2191 REM  *********************************
```

WYNN. Specimen results

4.29 Varying KS

It is easy to write a short program to set the T(N) equal to the partial sums of the series

$$1 - 1/2 + 1/3 - 1/4 + \dots \tag{4.54}$$

which represents $\ln 2 = 0.693\,147\,18$. Attaching this program to the front of WYNN, we analysed the resulting T(N) sequence. Taking 12 terms of the series yields only the information from the T(N) that ln 2 is somewhere between 0.737 and 0.653. However, printing out the approximants to the sequence gives the results shown below in table 4.4.

Table 4.4 Approximants for the $\ln(1 + x)$ series ($x = 1$).

	KS = 1	KS = 0	KS = −1
(1/1)	0.7	0.7	0.7
(2/2)	0.693 333 33	0.693 277 31	0.693 223 44
(3/3)	0.693 152 45	0.693 148 87	0.693 147 60
(4/4)	0.693 147 33	0.693 147 20	0.693 147 18
(5/5)	0.693 147 19	0.693 147 18	0.693 147 18

4.30 Perturbation series analysis

Killingbeck (1985a) gave some numerical results obtained by applying Wynn's algorithm for Padé approximants to the Euler series (§4.21). We give below some results obtained by using WYNN in conjunction with HYPOSC. The energy perturbation series produced by HYPOSC shows a rate of convergence which can be varied by changing a parameter K in the program. The series obtained with K = 0 are series of Stieltjes, and so it is to be anticipated that the approximant values will show the straddling properties explained in §4.21, with the exact perturbed energy playing the role of the $I(\lambda)$ in that discussion. However, it is at K = 0 that the series is strongly divergent, with the partial sums varying over many orders of magnitude, so that a check for rounding errors should be carried out using the variable scaling parameter SF (section 4.27).

The results given refer to the perturbed oscillator problem with the potential $x^2 + \lambda x^4$ and with a λ value of 5. To obtain a good ground-state energy of 2.018 340 7 simply by observing directly the partial sums of the series it is necessary to use a K value of 4 and take the series up to the λ^{20} term. The series for both K = 0 and K = 1 are both so quickly divergent that we can get no idea of the perturbed energy from the K = 0 series, while the partial sums of the K = 1 series show a semi-convergence to a value around 2.02. Tables 4.5 and 4.6 below show some test results for various approximants to the K = 0 and K = 1 series.

Table 4.5 Results for the K = 0 series, with KS = 1.

SF (M/N)	IE0	IE − 5	IE − 10
(7/7)	1.920 994 4	1.921 020 0	1.920 925 6
(8/7)	2.129 680 8	2.129 890 2	1.129 439 4
(8/8)	1.941 768 8	1.942 352 7	1.941 360 0
(9/8)	2.103 593 0	2.106 040 0	2.102 214 3
(9/9)	1.956 740 6	1.962 552 0	1.954 767 5
(10/9)	2.082 986 7	2.096 506 1	2.076 784 8

Table 4.6 Results for the K = 1 series, with KS = 1.

SF (M/N)	IE0	IE7	IE − 7
(7/7)	2.018 336 3	2.018 336 3	2.018 336 3
(8/7)	2.018 344 5	2.018 344 4	2.018 344 4
(8/8)	2.018 338 9	2.018 338 9	2.018 338 9
(9/8)	2.018 342 0	1.018 342 0	2.018 342 0
(9/9)	2.018 339 9	2.018 339 9	2.018 339 9
(10/9)	2.018 340 5	2.018 341 0	2.018 341 1

4.31 Comments on the results

The results in the tables were obtained on a 48K Spectrum microcomputer and other microcomputers would give slightly different results. For example, a double-precision arithmetic facility would cut down the amount of variation of the results in table 4.5 as SF varies. Nevertheless, it is clear that the uncertainty in the approximant values in table 4.5 is small compared with the error in the Padé estimates of the correct perturbed energy. Since the energy series at K = 0 is actually a series of Stieltjes type, the exact values of the approximants will give upper and lower bounds to the energy. Table 4.5 would thus allow us to obtain the estimate $1.95 < E < 2.10$ when we consider the uncertainty in the approximant values. A similar ground-state calculation for $\lambda = 1$ (rather than $\lambda = 5$) gives (at $K = 0$) the estimate $1.3919 < E < 1.3927$. The series for $K > 0$ have not so far been *proved* to be of Stieltjes type by mathematical analysis, although it has been observed empirically that the approximants do show the straddling property if K is not too large. Table 4.6 shows that the $K = 1$ series (which without Padé analysis gives the estimate $E \simeq 2.02$) seems to be converging to $E \simeq 2.018341$, with only a very small error in the approximant values.

5 The matrix inverse and generalized inverse

Programs

MATIN, GENFIT, INVERT.

5.1 General introduction

This chapter gives methods for the calculation of the inverse of a real square matrix (thus permitting the solution of systems of linear equations) and for the calculation of the generalized inverse of a rectangular real matrix (thus permitting the solution of least-squares fitting problems). In treating the solution of the least-squares polynomial fitting problem a slightly unusual approach has been taken. Instead of using the direct singular value decomposition method (Nash 1990), it has been decided to use a 'quick but dirty' way of producing the generalized inverse, together with the undeservedly neglected method of Schultz which will then refine the initial approximate generalized inverse. Apart from its value in this two-stage process the Schultz method will also produce an inverse or generalized inverse on its own, given the correct starting conditions; the program INVERT applies this approach.

MATIN and GENFIT. Mathematical theory

5.2 Choice of reduction method

The standard method of solving systems of linear equations is by using row or column transformations to reduce the system to triangular form (Gauss elimination) or to diagonal form (Jordan elimination). The second process is used in MATIN, since it also produces the inverse matrix. The elimination routine in MATIN is also used in GENFIT, which can handle few-term-polynomial least-squares fitting to a set of data points.

5.3 Least-squares fitting. The generalized inverse

The usual formal solution to the system of linear equations

$$Mx = y \tag{5.1}$$

is that involving the inverse matrix M^{-1},

$$y = M^{-1}x. \tag{5.2}$$

This solution is appropriate if the matrix M is square and non-singular. However, if M is rectangular, so that there are *more* equations than unknowns, an exact solution to the system (5.1) is in general impossible. A *least-squares* solution can be sought by forming the sum of squared residuals

$$\Delta = \sum_j \left[\sum_k M_{jk} x_k - y_j \right]^2 \tag{5.3}$$

and then seeking the set of x_k which gives it a minimum value. Differentiating (5.3) with respect to x_k and setting the result equal to zero for each k gives a set of equations which may be expressed in the matrix form

$$M^{\mathrm{T}} M x = M^{\mathrm{T}} y \tag{5.4}$$

where we assume all quantities to be real and denote by M^{T} the transpose of M. The result (5.4) also follows formally by multiplying (5.1) by M^{T} on both sides, although that procedure does not reveal the least-squares interpretation of the result. When M is rectangular, $M^{\mathrm{T}}M$ will be square. If $M^{\mathrm{T}}M$ is non-singular (as it is in most applications) equation (5.4) is a meaningful equation with the solution

$$x = (M^{\mathrm{T}}M)^{-1} M^{\mathrm{T}} y. \tag{5.5}$$

If M is square and non-singular then the equations (5.2) and (5.5) give the same results. For rectangular M, the matrix appearing in (5.5) is called the *generalized inverse* of M, in the common case that $M^{\mathrm{T}}M$ is non-singular. A generalized inverse *can* be defined in the more general case when $M^{\mathrm{T}}M$ is singular (Penrose 1955, Peters and Wilkinson 1970), but the case (5.5) corresponds to that arising most often in least-squares fitting calculations, where the equations (5.4) are called the normal equations.

5.4 An example. Parabolic fitting

The notation used in the theory of the generalized inverse is at odds with that usually employed in the theory of few-parameter least-squares fitting, where the x_k are often the given data point coordinates rather than the unknowns. For example, if a parabolic curve of form $Ax^2 + Bx + C$ is to be fitted to a set of data pairs (x_k, y_k), then the three parameters A, B and C correspond to the unknowns in the equations (5.4) and the elements of the matrix M are powers of the coordinates x_k. The formal equations (5.4) become

$$\begin{pmatrix} 1 & x_1 & x_1^2 \\ 1 & x_2 & x_2^2 \\ \vdots & \vdots & \vdots \end{pmatrix} \begin{pmatrix} C \\ B \\ A \end{pmatrix} = \begin{pmatrix} y_1 \\ y_2 \\ \vdots \end{pmatrix}. \tag{5.6}$$

Forming the matrix $M^{\mathrm{T}}M$ thus produces the normal equations

$$\begin{pmatrix} \Sigma 1 & \Sigma x & \Sigma x^2 \\ \Sigma x & \Sigma x^2 & \Sigma x^3 \\ \Sigma x^2 & \Sigma x^3 & \Sigma x^4 \end{pmatrix} \begin{pmatrix} C \\ B \\ A \end{pmatrix} = \begin{pmatrix} \Sigma y \\ \Sigma xy \\ \Sigma x^2 y \end{pmatrix} \tag{5.7}$$

in an obvious notation. To find C, B and A for a *single* set of data (x_k, y_k) it suffices to solve the equations (5.7). If several different sets of y_k are to be fitted, however, it may be worthwhile to find the generalized inverse first, so that the parameters C, B and A for each set of y_k can be obtained by simple matrix multiplication. Turing (1948) pointed out that one of the advantages of finding the matrix inverse when solving linear equations is that it is then possible to estimate how sensitive the x column is to small variations in the y column. For the case of a least-squares fitting calculation this is an important point; the use of too many unknown parameters can lead to normal equations which are ill-conditioned, although this problem rarely arises in few-parameter calculations.

5.5 An iterative method

Schultz (1933) proposed the iterative formula

$$X_{n+1} = 2X_n - X_n M X_n \tag{5.8}$$

for the calculation of the inverse M^{-1} of a square matrix M. If the input matrix X_n obeys the equation

$$MX_n = \mathbf{1} + \varepsilon \tag{5.9}$$

where **1** is the unit matrix and ε is an error matrix, then substitution in (5.8) leads to the result

$$MX_{n+1} = \mathbf{1} + \varepsilon^2. \tag{5.10}$$

Accordingly, if ε is a matrix sufficiently small for its increasing powers to tend to the null matrix, then the process (5.8) converges quadratically to M^{-1}. Duck (1964) gave an improved form of (5.8) which treats the upper and lower parts of X separately and which appears to have somewhat faster convergence. However, the most interesting development in the use of (5.8) was the discovery that when M is rectangular the iterative process can still be applied and will converge to the generalized inverse of M, provided that the initial X is a sufficiently small multiple of M^{T} (Ben-Israel 1966). The iterative calculation is slower than direct methods for inverting a non-singular matrix M, but the applicability of the iterative algorithm to rectangular matrices does give it an extra degree of generality. The program INVERT applies the Schultz iterative method to calculate a generalized inverse, starting from a multiple of M^{T}. The program GENFIT, however, uses the main modules of MATIN to find $(M^{\mathrm{T}}M)^{-1}$ fairly quickly and then produces the product $(M^{\mathrm{T}}M)^{-1}M^{\mathrm{T}}$ as its estimate of the generalized inverse. The effects of ill-conditioning and rounding error mean that the estimate will probably be slightly in error, as discussed by Nash (1990). The Schultz iterative module REFINER can then be applied to correct the generalized inverse; usually only one cycle of iteration is needed, whereas 20 or more would be needed if the Schultz process were used on its own.

MATIN and GENFIT. Programming notes

5.6 Program module structure

The program MATIN is designed to take an input square matrix M, compute $I = M^{-1}$ and then use I to solve the equation system $MX = Y$ for any input column Y. Although the solution of $MX = Y$ for a single Y is more speedily performed by reducing M down to triangular form, with a subsequent back substitution, we have chosen to form M^{-1} first. This takes longer initially, but allows any number of Y columns to be used subsequently in a speedy process. The method used is essentially a translation into BASIC of the simple textbook method which is used in a hand calculation. The matrix M and the unit matrix **1** are set up initially, and then the same sequence of row operations is carried out on them both. The aim is to reduce M to the unit matrix, which simultaneously converts the other matrix into M^{-1}. MATIN is not very sophisticated; it uses partial pivoting, which has been shown by Peters and Wilkinson (1975) to be usually adequate. It reduces M to diagonal form first, putting in at the end the necessary divisors to produce the unit matrix, since the determinant of M follows easily from the penultimate form of M. A copy C of M is actually used during the transformations, so that the input M can be retained.

The program GENFIT proceeds by solving the equations

$$M^{\mathrm{T}}MX = M^{\mathrm{T}}Y \tag{5.11}$$

for the least-squares problem. As explained in the earlier discussion this is only a safe procedure for a few-parameter fitting procedure, but the availability of the supposed inverse $(M^{\mathrm{T}}M)^{-1}$ helps in checking whether there is any possibility of danger from ill-conditioning. The point about this approach is that it can use the modules of MATIN which form the matrix inverse and which solve the system of equations. All that is needed is to form the square matrix $M^{\mathrm{T}}M$ associated with the input data points, invert the matrix and form the product $(M^{\mathrm{T}}M)^{-1}M^{\mathrm{T}}$, using REFINER to improve the result.

5.7 Forming $M^{\mathrm{T}}M$ for GENFIT

The elements of $A = M^{\mathrm{T}}M$ are, as equation (5.7) illustrates, sums of powers of the data coordinates x_N, with the general rule

$$A(J, K) = \sum_N x_N^{J+K-2}. \tag{5.12}$$

Since all the elements along each diagonal in the matrix have the same value, it is possible to calculate the sums (5.12) first and then use them to fill in the matrix elements. There is some degree of variety possible in the computation of the sums of powers. The powering operation ↑ tends to be comparatively slow on most microcomputer BASICS, and, since it uses a transformation to logarithms, will usually not work if the input number has a negative sign. This latter defect often causes unexpected halts during scientific calculations when the simple operation ↑ appears in a program, although the joint use of the ABS and SGN functions together with ↑ will remove the problem if the sign of the result can be predicted from the sign of the operand (as for integer powers). For low integer powers it is often preferable to use expressions, albeit more cumbersome, which use the explicit multiplication operator *. Although direct formation of $M^{\mathrm{T}}M$ using (5.12) was used in one version of GENFIT, the version shown here forms the matrix M first (using * rather than ↑) and then derives $M^{\mathrm{T}}M$ and other quantities by matrix multiplication. This means that the matrix M is available to form the sum

$$\sum_j \left| \sum_k M_{jk}x_k - y_j \right| \tag{5.13}$$

which gives a measure of the accuracy of the fitting process. M can also be used to form the product $(M^{\mathrm{T}}M)^{-1}M^{\mathrm{T}}$, which is the required generalized inverse.

5.8 MATIN. Program analysis and program

Lines 10 to 30 set up the matrix array M, the copy array C and the array I for the inverse matrix. Columns X and Y are for use in solving the equation system MX = Y.

Lines 100 to 140 are for the input of the matrix M and could be bypassed if a subroutine to set up M is available.

Lines 300 to 330 form a copy C of M, since the subsequent operations will modify the elements of C. A Boolean function is used in line 320 to set up the initial I matrix as the unit matrix.

Lines 400 to 480 look for the largest element in the bottom half of the current working column M (partial pivoting) and identify it as being in row L. The parity indicator PI is set equal to one in line 400, but is reversed in sign in line 480 if it turns out that two rows of the matrix are to be exchanged in the pivoting process (i.e. if $L \neq M$). PI is required because the sign of the determinant is reversed under interchange of two rows. If L = M then no interchange is needed and line 470 transfers control straight to the row combination module.

Lines 500 to 530 interchange rows M and L if necessary, using the single temporary store T.

Lines 600 to 680 form each new row J of *both* C and I by using the appropriate weighting factor R, which is worked out *once* in line 620 and then used several times inside the K loop. Line 610 ensures that the row M is left unchanged. It would also be possible to do this by omitting line 610 and incorporating a Boolean factor $(J \neq M)$ in the definition of R. Note that the M loop runs all the way from line 410 to line 680.

Lines 700 to 760 scale up or down the elements of I as required to ensure that C becomes the unit matrix and I becomes the inverse M^{-1}. However, C is *not* converted explicitly to unit matrix form; its diagonal elements are multiplied together (with the parity indicator from line 700) to give the value of the determinant of M.

Lines 800 to 890 accept any input column Y and print out the solution X = IY to the equation system MX = Y. The X and Y columns are stored, but the manual command GO TO 800 will permit the use of a new Y column.

Lines 900 to 980 form an error sum SS to indicate how well the product

MX actually approximates Y. In principle (although it is hardly necessary) the module REFINER could be added at the end of the program to improve the inverse I and diminish SS. Line 950, which exhibits both the actual and the calculated Y column, is usually remmed out of action, since the error is negligible for matrices of reasonable size which have elements not showing a large range of orders of magnitude.

```
  7 REM  ********************************
  8 REM  MATIN
  9 REM  ********************************
 10 INPUT "DIM";N
 20 DIM A(N,N): DIM C(N,N): DIM I(N,N)
 30 DIM X(N): DIM Y(N)
 99 REM  ********************************
100 PRINT "INPUT MATRIX A"
101 REM  ********************************
110 FOR J=1 TO N: FOR K=1 TO N
120 PRINT J;K: INPUT X
130 PRINT " ";X: LET A(J,K)=X
140 NEXT K: NEXT J
296 REM  ********************************
297 REM  SOLVER
298 REM  COPY A,SET UP I
299 REM  ********************************
300 FOR J=1 TO N: FOR K=1 TO N
310 LET C(J,K)=A(J,K)
320 LET I(J,K)=(J=K)
330 NEXT K: NEXT J
397 REM  ********************************
398 REM  FIND LARGEST PIVOT
399 REM  ********************************
400 LET PI=1
410 FOR M=1 TO N: LET U=0
420   FOR I=M TO N
430 LET V=ABS C(I,M)
440 IF V<=U THEN GO TO 460
450 LET U=V: LET L=I
460   NEXT I
465 REM PRINT "M,L";M;L
470 IF L=M THEN GO TO 600
480 LET PI=-PI
497 REM  ********************************
498 REM  EXCHANGE ROWS
499 REM  ********************************
500 FOR K=1 TO N
510 FOR T=C(L,K): LET C(L,K)=C(M,K): LET C(M,K)=T
```

```
520 LET T=I(L,K): LET I(L,K)=I(M,K): LET I(M,K)=T
530 NEXT K
597 REM  ***********************************
598 REM  COMBINE ROWS
599 REM  ***********************************
600 FOR J=1 TO N
610 IF J=M THEN GO TO 670
620 LET R=C(J,M)/C(M,M)
630    FOR K=1 TO N
640 LET C(J,K)=C(J,K)-R*C(M,K)
650 LET I(J,K)=I(J,K)-R*I(M,K)
660    NEXT K
670 NEXT J
680 NEXT M
697 REM  ***********************************
698 REM  DET AND INVERSE
699 REM  ***********************************
700 LET DET=PI
710 FOR J=1 TO N: LET D=C(J,J)
720 LET DET=DET*D
730    FOR K=1 TO N
740 LET I(J,K)=I(J,K)/D
750 NEXT K: NEXT J
760 PRINT "DET";DET
797 REM  ***********************************
798 REM  INPUT Y,FIND X
799 REM  ***********************************
800 PRINT "INPUT Y COLUMN"
810 FOR J=1 TO N: PRINT J
820 INPUT X: PRINT " ";X
830 LET Y(J)=X
840 NEXT J: PRINT
841 REM  ***********************************
850 FOR J=1 TO N: LET S=0
860    FOR K=1 TO N
870 LET S=S+I(J,K)*Y(K): NEXT K
880 PRINT S: LET X(J)=S
890    NEXT J: PRINT
897 REM  ***********************************
898 REM  FORM ERROR SUM
899 REM  ***********************************
900 LET SS=0: FOR J=1 TO N
910 LET S=0: FOR K=1 TO N
920 LET S=S+A(J,K)*X(K)
930 NEXT K
940 LET SS=SS+ABS (S-Y(J))
950 PRINT Y(J),S
960 NEXT J
```

```
970 PRINT "ERROR";SS
980 STOP
981 REM  ********************************
```

5.9 GENFIT. Program analysis and program

The central SOLVER section of GENFIT is taken directly from the program MATIN in its original form. Other modules from SOLVER are also used with only slight changes in the naming of the loop parameters and variables.

Lines 10 to 50 set up the required arrays in terms of the number ND of data points and the maximum number NP of parameters to be used in the fitting process. G is used for the generalized inverse IM^T and I for the ordinary inverse $(M^TM)^{-1}$. S and T are temporary matrices needed during the Schultz iteration process in REFINER.

Lines 100 to 160 set up the matrix M required in the least-squares polynomial fitting process. The loop between lines 140 and 160 builds up the required powers of x_J by using the $*$ operation rather than the $\uparrow$ operation.

Lines 200 to 260 form the matrix $A = M^TM$. Note the use of M(L, J), rather than M(J, L), in line 230 to ensure that the left-hand factor is the transpose M^T.

Line 290 calls for the choice of a number N of parameters. N can be less than or equal to the NP value chosen in line 20. The rest of the computation uses appropriate submatrices of the full M and A matrices, producing a least-squares fit to an N-term polynomial. To vary N the manual command GO TO 290 or GO TO 200 can be used.

Lines 300 to 770 are the standard modules from MATIN. They form the matrix inverse $I = A^{-1} = (M^TM)^{-1}$. The GO TO 1000 at line 770 then transfers control to a later module.

Lines 800 to 980 are, apart from using ND as upper limit in some loops and using G instead of I in line 870, the standard MATIN modules for receiving the data column Y and producing the output column X, in this case a column of best-fit parameters. The error sum SS is displayed to indicate how the fit improves as N is increased or as REFINER improves the parameters.

Lines 1000 to 1070 form the matrix product IM^T to give the generalized inverse G of M. Note the use of M(K, L), instead of M(L, K), in line 1030 to give M^T rather than M as the right-hand factor in the matrix product.

Line 1070 has a STOP command. At this point the matrix G has been formed. The manual command GO TO 800 allows the Y data to be inserted and displays the best-fit parameters and error sum. GO TO 1200 then refines G by a Schultz iteration and displays the revised solution. The reader will quickly see how the use of these two commands allows multiple refinement of G, the use of new Y data, etc, in a variety of combinations. GO TO 200 allows the adjustment of the number N of parameters used in the fitting process.

Lines 1200 to 1360 constitute REFINER, which takes in an approximate generalized inverse G of M and produces an improved G by applying one cycle of Schultz iteration (although it could be put inside a loop to perform two or more cycles). The first step produces $T = \mathbf{1} - MG$, the second step produces $S = GT$. S is thus equal to $G - GMG$, and so is the correction matrix which must be added to G to produce the new estimate $2G - GMG$ which is required. Note the use of a Boolean function in line 1250 to form $\mathbf{1} - MG$ and the adding of the correction to G in line 1310, with S being copied into G in line 1340. The GO TO 850 statement in line 1360 ensures that the analysis of the stored Y data is carried out at once without a request for further input. A GO TO 800 could be used instead if desired.

To use GENFIT to find the least-squares solution for an arbitrary matrix M it is clearly sufficient to modify the input routine to accept the elements of M directly.

```
  7 REM  *********************************
  8 REM  GENFIT
  9 REM  *********************************
 10 INPUT "NO.OF DATA PTS";ND
 20 INPUT "NO.OF PARAMETERS";NP
 30 DIM M(ND,NP): DIM G(NP,ND)
 40 DIM A(NP,NP): DIM C(NP,NP): DIM I(NP,NP)
 50 DIM T(ND,ND): DIM S(NP,ND)
 60 DIM X(NP): DIM Y(ND)
 99 REM  *********************************
100 PRINT "INPUT X VALUES"
101 REM FORM M MATRIX
102 REM  *********************************
110 FOR J=1 TO ND
120 PRINT J: INPUT X
130 PRINT " ";X: LET F=1
140 FOR K=1 TO NP
150 LET M(J,K)=F: LET F=F*X
160 NEXT K: NEXT J
197 REM  *********************************
198 REM FORM A=MT*M
199 REM  *********************************
200 FOR J=1 TO NP
210 FOR K=1 TO NP: LET S=0
```

```
220 FOR L=1 TO ND
230 LET S=S+M(L,J)*M(L,K)
240 NEXT L
250 LET A(J,K)=S
260 NEXT K: NEXT J
289 REM  ********************************
290 INPUT "NO.OF PARAMETERS";N
291 REM  ********************************
297 REM  SOLVER
298 REM COPY A,SET UP I
299 REM  ********************************
300 FOR J=1 TO N: FOR K=1 TO N
310 LET C(J,K)=A(J,K)
320 LET I(J,K)=(J=K)
330 NEXT K: NEXT J
397 REM  ********************************
398 REM  FIND LARGEST PIVOT
399 REM  ********************************
400 LET PI=1
410 FOR M=1 TO N: LET U=0
420   FOR I=M TO N
430 LET V=ABS C(I,M)
440 IF V<=U THEN GO TO 460
450 LET U=V: LET L=I
460   NEXT I
465 REM PRINT "M,L";M;L
470 IF L=M THEN GO TO 600
480 LET PI=-PI
497 REM  ********************************
498 REM  EXCHANGE ROWS
499 REM  ********************************
500 FOR K=1 TO M
510 LET T=C(L,K): LET C(L,K)=C(M,K): LET C(M,K)=T
520 LET T=I(L,K): LET I(L,K)=I(M,K): LET I(M,K)=T
530 NEXT K
597 REM  ********************************
598 REM  COMBINE ROWS
599 REM  ********************************
600 FOR J=1 TO N
610 IF J=M THEN GO TO 670
620 LET R=C(J,M)/C(M,M)
630   FOR K=1 TO N
640 LET C(J,K)=C(J,K)-R*C(M,K)
650 LET I(J,K)=I(J,K)-R*(M,K)
660   NEXT K
670 NEXT J
680 NEXT M
697 REM  ********************************
```

```
 698 REM   DET AND INVERSE
 699 REM   **********************************
 700 LET DET=PI
 710 FOR J=1 TO N: LET D=C(J,J)
 720 LET DET=DET*D
 730    FOR K=1 TO N
 740 LET I(J,K)=I(J,K)/D
 750 NEXT K: NEXT J
 760 PRINT "DET";DET
 770 GO TO 1000
 797 REM   **********************************
 798 REM   INPUT Y,FIND X
 799 REM   **********************************
 800 PRINT "INPUT Y COLUMN"
 810 FOR J=1 TO ND: PRINT J
 820 INPUT X: PRINT " ";X
 830 LET Y(J)=X
 840 NEXT J: PRINT
 841 REM   **********************************
 850 FOR J=1 TO N: LET S=0
 860    FOR K=1 TO ND
 870 LET S=S+G(J,K)*Y(K): NEXT K
 880 PRINT S: LET X(J)=S
 890 NEXT J: PRINT
 897 REM   **********************************
 898 REM   FORM ERROR SUM
 899 REM   **********************************
 900 LET SS=0: FOR J=1 TO ND
 910 LET S=0: FOR K=1 TO N
 920 LET S=S+M(J,K)*X(K)
 930 NEXT K
 940 LET SS=SS+ABS (S-Y(J))
 950 REM PRINT Y(J),S
 960 NEXT J
 970 PRINT "ERROR";SS
 980 STOP
 997 REM   **********************************
 998 REM   FORM G=I*MT
 999 REM   **********************************
1000 FOR J=1 TO N
1010 FOR K=1 TO ND: LET S=0
1020 FOR L=1 TO N
1030 LET S=S*I(J,L)*M(K,L)
1040 NEXT L
1050 LET G(J,K)=S
1060 NEXT K: NEXT J
1070 STOP
1196 REM   **********************************
```

```
1197 REM  REFINER
1198 REM  FORM T=1-M*G
1199 REM  ********************************
1200 FOR J=1 TO ND
1210 FOR K=1 TO ND: LET S=0
1220 FOR L=1 TO N
130 LET S=S+M(J,L)*G(L,K)
1240 NEXT L
1250 LET T(J,K)=(J=K)-S
1260 NEXT K: NEXT J
1267 REM  ********************************
1268 REM  FORM S=G+G*T
1269 REM  ********************************
1270 FOR J=1 TO N: FOR K=1 TO ND
1280 LET S=0: FOR L=1 TO ND
1290 LET S=S+G(J,L)*T(L,K)
1300 NEXT L
1310 LET S(J,K)=G(J,K)+S
1320 NEXT K: NEXT J
1329 REM  ********************************
1330 FOR J=1 TO N: FOR K=1 TO ND
1340 LET G(J,K)=S(J,K)
1350 NEXT K: NEXT J
1360 GO TO 850
1361 REM  ********************************
```

GENFIT. Specimen results

5.10 A parallel test

To test the program the results from a sequence of midpoint integrations of the function $(1 + x^2)^{-1}$ were used. The limits of integration were 0 and 1. From the usual principles of Romberg integration we know that the error law should be of (h^2, h^4) type, and the data actually give the result 0.785 398 17 in a Romberg integration approach. The idea of the test, however, is to let the fitting program try to fit the data to polynomials of increasing order in h and see what happens. The data used are shown in table 5.1.

Only the last five digits of the numbers were used as the (integer) y values, since the first three digits are common to all the data. The quantities 100/NS were used as the input x_N values for the data points. Table 5.2 shows the rounded results, with the parameters first, followed by the error sum in parentheses.

The parameter results in both columns agree quite well throughout the table. At N = 3 the decrease in the fitting error is dramatic and the results show clearly the presence of the strong h^2 term in the data. The

Table 5.1 Midpoint integration data.

NS	I
20	(0.785)450 25
40	411 18
80	401 42
160	398 98

Table 5.2 GENFIT results with and without REFINER.

N	With	Without
2	38 719.5	38 719.5
	1205.9	1205.9
	(1402.0)	(1402.0)
3	39 817.2	39 817.2
	−0.8439	−0.8442
	208.48	208.48
	(0.2902)	(0.2890)
4	39 816.5	39 816.5
	0.4003	0.4399
	207.89	207.89
	0.0732	0.0748
	(0.0022)	(0.3635)

h term, required in the straight-line fit at $N = 2$, acquires a relatively negligible coefficient at $N = 3$. At $N = 4$ the relative smallness of the h and h^3 terms is clear, with the constant and h^2 terms dominating. The estimate of I at both $N = 3$ and $N = 4$ is 39 817, the correct value, and the h^2 coefficient is given as 208 (for the x units chosen). The only problematic feature about the results is the incorrect *increase* of the unrefined error on passing from $N = 3$ to $N = 4$. The value of the determinant DET is much less at $N = 4$, suggesting some difficulty with ill-conditioning. The refined results behaves better, producing a fair attempt at the zero error which should theoretically be obtained at $N = 4$. The results obtained by using the slower program INVERT to treat the problem are, of course, almost identical to those obtained by using the REFINER module in GENFIT. If REFINER is applied repeatedly, using GO TO 1200, then the results show a slight fluctuation due to residual rounding errors; the magnitude of this depends on the number of digits accuracy available on the microcomputer being used.

The results of the test calculation show that the direct solution of the normal equations, although it fits easily into the traditional method for solving linear equations, leads to results which may need further refinement as the number of parameters is increased. However, it must be admitted that the example was deliberately treated in a manner intended to increase the errors, so that the effect of the REFINER module could be made explicit. One of the obvious ways to decrease problems due to the ill-conditioning of the matrices is to use the average of the x_N data values as the x_N origin. Since this will produce some negative coordinate values in the computation, it is important that any powering operations should use the BASIC operation * rather than ↑. The x_N coordinates 100/NS were used as input to get the results discussed above. The y coordinates were treated sensibly, by converting them from eight-digit floating-point form to five-digit integer form, which was made possible by the special properties of the data. If the x coordinates used are also adjusted, by subtracting from them the fixed number 3 (their approximate mean), then the fit to the data is much improved and the REFINER corrections become less important. The error sum is reduced to a value of order 1E − 4 at N = 4 on a typical microcomputer which computes with roughly nine-decimal-digit accuracy. Since the variable $x - 3$ rather than x is being used, the first three coefficients are different from those obtained in the previous calculation, but the sensitive fourth coefficient (which theoretically should agree in both calculations) is obtained as 0.073 13, very close to its value in the original calculation.

5.11 INVERT. Program analysis and program

Most of the sections of INVERT have already been used in GENFIT. The REFINER module, however, has to be used repeatedly in INVERT and so is regarded as a subroutine. This allows it to be kept at line 1200, making it easier to put INVERT together initially from prerecorded modules. The modules for finding the best-fit parameters are also kept exactly as in GENFIT.

Lines 20 and 60 set up the required arrays as in GENFIT, except that A, C, and I arrays are not needed.

Lines 100 to 160 take in the x data values and form the matrix M as in GENFIT. The only change is that the sum T is formed. This equals the diagonal sum of the matrix MM^T, and is needed later.

Lines 200 to 240 allow a number of parameters N less than or equal to NP to be chosen for the calculation, and then set up a multiple of the transpose M^T as the initial estimate of the generalized inverse of M. The scaling factor SF which divides M^T must, according to Ben-Israel (1966), be greater than T/2, where T is the sum evaluated during the input stage. The iterative process will then converge. We have found that for the special case N = 1 an even greater value is required, and this fact is incorporated in the expression in line 210.

Lines 250 to 290 control the number of iterations NI to be performed. If NI is set equal to zero then the rest of the calculation proceeds. The subroutine 1200 is the REFINER module which performs the Schultz iteration.

Lines 800 to 980 are the modules from GENFIT which work out GY to get the best-fit parameters and also show the error sum which measures the closeness of fit.

Lines 1200 to 1360 constitute the REFINER subroutine. The only important changes from GENFIT are in line 1370, where RETURN is added to make the module a subroutine, and in lines 1260, 1300 and 1320. These lines use a variable SS which measures the magnitude of the adjustment being made to G on each cycle. SS increases through a maximum and then decreases to zero, with typically 10 to 40 cycles being required for G to converge. The operator monitors the displayed S values and chooses NI in line 250 accordingly each time an input for NI is required. The process could, of course, be made automatic by a slight complication of the program, but we have chosen a displayed form which is appropriate for interactive computation on a microcomputer.

```
  7 REM  *********************************
  8 REM  INVERT
  9 REM  *********************************
 10 INPUT "NO.OF DATA PTS";ND
 20 INPUT "NO.OF PARAMETERS";NP
 30 DIM M(ND,NP): DIM G(NP,ND)
 50 DIM T(ND,ND): DIM S(NP,ND)
 60 DIM X(NP): DIM Y(ND)
 99 REM  *********************************
100 PRINT "INPUT X VALUES"
101 REM  FORM M MATRIX
102 REM  *********************************
105 LET T=0
110 FOR J=1 TO ND
120 PRINT J: INPUT X
130 PRINT " ";X: LET F=1
140 FOR K=1 TO NP
150 LET M(J,K)=F: LET T=T+F*F: LET F=F*X
160 NEXT K: NEXT J
197 REM  *********************************
```

```
 198 REM  SET G=MT/SF
 199 REM  ********************************
 200 INPUT "NO.OF PARAMETERS";N
 210 LET SF=T/(2-(N=1))
 220 FOR J=1 TO N: FOR K=1 TO ND
 230 LET G(J,K)=M(K,J)/SF
 240 NEXT K: NEXT J
 249 REM  ********************************
 250 INPUT "NO.OF ITERATIONS";NI
 251 REM  ********************************
 260 IF NI=0 THEN GO TO 800
 270 FOR I=1 TO NI
 280 GO SUB 1200
 290 NEXT I: GO TO 250
 797 REM  ********************************
 798 REM  INPUT Y,FIND X
 799 REM  ********************************
 800 PRINT "INPUT Y COLUMN"
 810 FOR J=1 TO ND: PRINT J
 820 INPUT X: PRINT " ";X
 830 LET Y(J)=X
 840 NEXT J: PRINT
 841 REM  ********************************
 850 FOR J=1 TO N: LET S=0
 860   FOR K=1 TO ND
 870 LET S=S+G(J,K)*Y(K): NEXT K
 880 PRINT S: LET X(J)=S
 890 NEXT J: PRINT
 897 REM  ********************************
 898 REM  FORM ERROR SUM
 899 REM  ********************************
 900 LET SS=0: FOR J=1 TO ND
 910 LET S=0: FOR K=1 TO N
 920 LET S=S+M(J,K)*X(K)
 930 NEXT K
 940 LET SS=SS+ABS (S-Y(J))
 950 PRINT Y(J),S
 960 NEXT J
 970 PRINT "ERROR";SS
 980 STOP
 997 REM  ********************************
1197 REM  REFINER
1198 REM  FORM T=1-M*G
1199 REM  ********************************
1200 FOR J=1 TO ND
1210 FOR K=1 TO ND: LET S=0
1220 FOR L=1 TO N
1230 LET S=S*M(J,L)*G(L,K)
```

```
1240 NEXT L
1250 LET T(J,K)=(J=K)-S
1260 NEXT K: NEXT J: LET SS=0
1267 REM  **********************************
1268 REM  FORM S=G+G*T
1269 REM  **********************************
1270 FOR J=1 TO N: FOR K=1 TO ND
1280 LET S=0: FOR L=1 TO ND
1290 LET S=S+G(J,L)*T(L,K)
1300 NEXT L: LET SS=SS+ABS S
1310 LET S(J,K)=G(J,K)+S
1320 NEXT K: NEXT J: PRINT SS
1329 REM  **********************************
1330 FOR J=1 TO N: FOR K=1 TO ND
1340 LET G(J,K)=S(J,K)
1350 REM PRINT G(J,K)
1355 POKE 23592,10
1360 NEXT K: NEXT J
1370 RETURN
1371 REM  **********************************
```

6 The matrix eigenvalue problem

Programs

JOLLY, FOLDER, HITTER, BETOSC.

6.1 General introduction

This chapter describes three methods for the calculation of matrix eigenvalues and eigencolumns. The method JOLLY is of recent origin and so its characteristics have not been fully explored. It is included here because of the very simple algebra which lies at the heart of the algorithm and which makes possible the speedy eigenvalue deflation method of §6.6. The program FOLDER is the one most used by the author in numerical work. It will work for the generalized eigenvalue problem and is designed to fit into the root-finding approach outlined in chapter 2. The description of FOLDER given here brings together background material from perturbation theory and the theory of linear equations to help to clarify the principles on which it is based. The program HITTER belongs to a large family of iterative methods for the eigenvalue problem, and has the special advantage of being applicable to very large matrix problems in which the matrix elements are not stored numerically but are evaluated analytically when required. This feature is exploited later in the two problems of chapter 10. The ancillary program BETOSC of this chapter provides a quantum mechanical test matrix which could be used in conjunction with any eigenvalue program.

JOLLY. Mathematical theory

6.2 The origin of JOLLY

The method of matrix eigenvalue calculation described here has been developed from one which was first described by Jolicard and Grosjean

(1985) in terms of a formalism based on perturbation theory and an adiabatic switching process. The present author has studied the method within the context of traditional matrix algebra, concluding that it represents probably the most simple and elegant transformation method in existence for finding (one at a time) the eigenvalues and eigencolumns of a square matrix. The next most simple method seems to be that of Jahn (1948), as refined by Collar (1948), although that, like the commonly used Jacobi method (Rutishauser 1966), is designed to find all the eigenvalues simultaneously. If the square $N \times N$ matrix is real but not symmetric then it may have less than N real eigenvalues. The real eigenvalues might still be calculable by a method which uses real arithmetic to find only one at a time, whereas a method designed to diagonalize the entire matrix at once would need to allow for the use of complex number arithmetic. The situation is similar to that with regard to finding the roots of an ordinary polynomial equation; indeed, a polynomial equation is precisely what is obtained from the secular equation associated with the matrix eigenvalue problem. Recently Jolicard and Perrin (1989) have briefly outlined a matrix-theoretic version of the method which was originally derived using adiabatic switching theory. Although the program and theory given here contain several novel features, the program has been given the short name JOLLY to acknowledge the original work of Georges Jolicard in setting out the basic principles which have been used.

6.3 Some simple transformations

We consider an $N \times N$ matrix H for which we wish to find an eigenvalue and eigenvector. From the basic defining equation

$$Hx = Ex \tag{6.1}$$

we can quickly establish the following property of the column Sx, where S is any non-singular $N \times N$ matrix:

$$(SHS^{-1})Sx = SHx = ESx. \tag{6.2}$$

This result means that the transformed matrix SHS^{-1} has the same eigenvalues as H, but with the eigencolumn Sx (instead of x) associated with the eigenvalue E.

The method to be described here uses $N \times N$ matrices of the form $\mathbf{1} + R(K)$, where $\mathbf{1}$ is the unit matrix. $R(K)$ is a null matrix except for the column K, which is only required to have the diagonal element $R(K, K)$ zero. If $R(K)$ and $R'(K)$ are two such matrices, then we obtain by explicit matrix multiplication the result

$$(\mathbf{1} + R)(\mathbf{1} + R') = \mathbf{1} + R + R' \tag{6.3}$$

where the index K has been omitted for the sake of brevity. The result (6.3) shows that a simple addition rule can be used when multiplying together matrices of the form $\mathbf{1} + R(K)$; it also shows that the result

$$(\mathbf{1} + R)^{-1} = \mathbf{1} - R \tag{6.4}$$

holds. This latter result makes the use of transformations of the type $S^{-1}HS$ particularly simple if S is taken in the form $S = \mathbf{1} + R(K)$, since the calculation of the inverse of S becomes trivial.

Suppose now that the Kth column of $S = \mathbf{1} + R(K)$ is actually an eigencolumn of H with eigenvalue E. By writing down an example and working out the triple product $(\mathbf{1} - R)H(\mathbf{1} + R)$ the reader may confirm that the following result holds: the transformed matrix $S^{-1}HS$ has a (K, K) element equal to E, with all other elements in the Kth column zero.

6.4 The algorithm

The aim of the method described here is to perform a sequence of transformations, each with $S = \mathbf{1} + R(K)$, until H is transformed to the reduced form described in the previous section. In that form an eigenvalue E is in the (K, K) position, with zeros elsewhere in the Kth column. By checking through an example again the reader may confirm that the transformation can be carried out using the following algorithm, in which $R(N, K)$ denotes the Nth element in the Kth column of $R(K)$. Note that step 1 modifies only a single row.

Step 1 (all M):

$$H(M, K) := H(M, K) + \sum_{N \neq K} H(M, N)R(N, K). \tag{6.5}$$

[Step 1 forms $H(\mathbf{1} + R)$.]

Step 2 (all N, M, excluding $M = K$):

$$H(M, N) := H(M, N) - H(K, N)R(M, K). \tag{6.6}$$

[Step 2 forms $(\mathbf{1} - R)H(\mathbf{1} + R)$.]

The two assignment statements above will carry out the transformation $S^{-1}HS$, but we still have to produce some rule to give the elements $R(N, K)$. The rule which seems most obvious in the case when the diagonal elements of the matrix H are different and the off-diagonal elements are small is the rule suggested by first-order perturbation theory:

$$R(N, K) = [H(K, K) - H(N, N)]^{-1}H(N, K) \tag{6.7}$$

for $N \neq K$. $R(K, K)$ is held at zero, so that $S(N, K)$ is one. This rule was used by Jolicard and Grosjean (1985) and they described the whole method as a perturbation method. Strictly speaking, they were only using a perturbative approach for prescribing the $R(N, K)$. The transformation equations described in the preceding sections are valid in general; the matrices $S^{-1}HS$ and H have the same eigenvalues for *any* non-singular S, although convergence to the special reduced form of $S^{-1}HS$ is the most simple way to obtain an eigenvalue.

When the matrix H is not diagonally dominant, or when the diagonal elements are not well separated, the formula (6.7) will considerably overestimate the elements $R(N, K)$, which are ideally intended to be the elements of an eigencolumn. There are several ways to prevent this overestimation by imposing some form of cut-off or attenuation. The simple approach adopted in JOLLY is to work out $R(N, K)$ according to the perturbation rule (6.7) and, calling it R, to replace it by the modified quantity

$$R' = R(\mathbf{1} + \beta|R|)^{-1} \tag{6.8}$$

where β can be a number (typically 1) or a Boolean function which 'turns on' above some critical R value.

6.5 The eigencolumns

When the transformed matrix is in the reduced form it has as eigencolumn (with eigenvalue E) the unit column which has a one in the Kth position and zeros elsewhere; we can denote this column $e(K)$. To reach this simple eigencolumn we have applied a sequence of transformations which combine according to the rule

$$(1 + R_1)(1 + R_2)(1 + R_3) = 1 + R_1 + R_2 + R_3. \tag{6.9}$$

It follows that to compute the eigencolumn for the original matrix H it is sufficient to retain a column which gives the total of the appropriate $R(N, K)$ elements so far. The requisite rule is as follows,

$$X(K) = 1 \tag{6.10}$$

$$(N \neq K) \qquad X(N) = \sum_J R_J(N, K) \tag{6.11}$$

where the J sum goes over the sequence of transformations used. When H has been transformed to the reduced form, with the eigenvalue E in the (K, K) position, any subsequent transformations will have zero $R(N, K)$ values for $N \neq K$. This means that $X(N)$ as given by (6.11) will remain unchanged, that is the sequence $X(N)$ will have converged to the eigencolumn associated with E. The column has the special normalization that $X(K) = 1$, although scaling can be applied if required.

6.6 Eigenvalues by deflation

When one eigenvalue and its eigencolumn have been found for an $N \times N$ matrix a process of deflation is often recommended. This process produces a matrix of dimension $(N-1) \times (N-1)$, from which the remaining $N-1$ eigenvalues are found. One commonly used form of deflation involves the calculated eigencolumn x in the combination xx^{T}, but there is another type of deflation which is closely related to the methods used here. In that method the known eigencolumn x is used in a transformation S which takes the matrix H over into exactly the same reduced form as that produced by the transformations described in the preceding sections. Close inspection of the relevant texts (Faddeeva 1959, Goult *et al* 1974) shows that the transforming matrix which they use is just the same as that arising in the present calculation. The further step which we have taken here (following Jolicard) is to base the *calculation* of the eigencolumns on the *same* kind of transformation. This possibility was not noted by the authors cited above. They supposed that the eigencolumns were to be obtained by some other technique, with the simple $\mathbf{1} + R$ type of transformation only being used in the deflation process.

The basic principle of the deflation process is best seen by inspecting a 3×3 case in detail; this will provide the relevant formulae, which we can then see how to generalize to the $N \times N$ case. Suppose that we have made the choice $K = 1$ in transforming the 3×3 matrix, so that at the end of the process the matrix has a reduced form, with a $(1, 1)$ element equal to the eigenvalue E_1 and with zeros in the rest of the first column. By the general principles explained previously, this reduced matrix has the same eigenvalues as the original matrix H from which it was derived. The eigenvalue problem for another eigenvalue E thus can be written in the form

$$\begin{pmatrix} E_1 & * & * \\ 0 & \multicolumn{2}{c}{\multirow{2}{*}{B}} \\ 0 & & \end{pmatrix} \begin{pmatrix} a \\ b \\ c \end{pmatrix} = E_2 \begin{pmatrix} a \\ b \\ c \end{pmatrix} \tag{6.12}$$

where $*$ denotes non-zero elements and B is a 2×2 submatrix. The first point to note is that the *eigenvalues* of B are also eigenvalues of the 3×3 matrix and so of H. This can be seen by noting that the determinant of any 3×3 matrix of this form has the determinant $|B|$ as a factor. This is due to the column of zeros which appears in the reduced form of the matrix, and shows that the algorithm used in JOLLY is naturally adapted to the deflation process. It is clear from inspection of (6.12) that the column (b, c) *must* be an eigencolumn of B corresponding to the eigenvalue E_2. This means that to find E_2 we simply apply our technique to the submatrix B and produce both E_2 and the eigencolumn (b, c). In that calculation the top row of the transformed 3×3 matrix plays no part. That row is only needed if we wish to find the $(1, 1)$ element a which augments (b, c) to give an eigencolumn of three elements associated with the newly found eigenvalue E_2. Calling the 3×3 matrix M (since it is *not* H, but a transformed form of H), we can write down the $(1, 1)$ element of the eigenvalue equation (6.12) to obtain the result

$$(E_2 - E_1)a = M_{12}b + M_{13}c. \tag{6.13}$$

This gives us the missing element a by making use of the elements of the first row of the matrix M. To convert the column (a, b, c) into an eigencolumn of the original matrix H, we have to multiply it by the S^{-1} matrix (i.e. $\mathbf{1} - R$) which was used in transforming H to M. The generalization of (6.13) for the deflation $N \to N - 1$ (instead of $3 \to 2$) is clearly

$$(E_2 - E_1)X(N) = \sum_{J<N} M(N, J)X(J) \tag{6.14}$$

with the $N - 1$ eigencolumn elements $X(J)$ having been found from the calculation of the submatrix which is the analogue of B in our specimen case.

6.7 Rapid eigenvalue calculation

Although the calculation described in the preceding section can be used fairly easily for the deflation $N \to N - 1$, continuing the process to $N - 2$, $N - 3$, etc has the difficulty that the total product of the transformations previously used is needed. For example, at the $N - 2$ stage the product of $\mathbf{1} - R(1)$ and $\mathbf{1} - R(2)$ will be needed to transform the calculated eigencolumn back into an eigencolumn of the original matrix H. It was decided not to incorporate this refinement in JOLLY, so as to cut down on the storage space required and on the complexity of the program.

The direct way to calculate E_1, E_2, etc is to vary the index K and stick to the full $N \times N$ matrix calculation; this method yields the associated eigenvectors directly. If only the *eigenvalues* are required, however, JOLLY can still exploit the deflation process to speed up the calculation, with no need to store details of previous transformations.

First E_1 is calculated for H; then E_2 is calculated from the resulting $(N-1) \times (N-1)$ submatrix; then E_3 is calculated from the resulting $(N-2) \times (N-2)$ submatrix; and so on. At the 2×2 stage, the last two eigenvalues emerge together, since the reduced form of a 2×2 matrix necessarily has the two matrix eigenvalues in the diagonal positions. Exploiting this sequential deflation process speeds up the eigenvalue calculation considerably. At the end of the calculation the transformed matrix has the N eigenvalues stored along the diagonal; it is not diagonalized in the traditional sense, since it has non-zero elements above the diagonal.

6.8 Non-symmetric matrices

Some of the traditional matrix eigenvalue techniques, such as the Jacobi one, are specifically designed for use with real symmetric matrices, which give only real eigenvalues. It should be quite obvious that JOLLY, although it uses real arithmetic (and so could not handle complex eigenvalues or eigenvectors), can work with non-symmetric real matrices; indeed, it specifically renders the starting matrix H non-symmetric, while leaving the eigenvalues unchanged. It follows, then, that the algorithm will work even if the starting matrix is not symmetric. In that case, the $N \times N$ matrix can have less than N real eigenvalues and so the question arises of whether the real arithmetic used in JOLLY is adequate to find all of those real eigenvalues. Because of the comparative newness of the method that question (and several others) has not yet been explored in detail, but we give a test example in the specimen results.

6.9 Preliminary transformations

Although the elements of $R(K)$ are given by the perturbation formulae (6.7) and (6.8) during the sequence of transformations which take H into the reduced form, it is clear from the mathematical theory that any $R(K)$ could be used. In particular, if an approximate eigencolumn is already known, in a form with $X(K) = 1$, then it can be used in a preliminary transformation to bring H close to its reduced form. The appropriate elements must be added in to the sum (6.11) in order to keep track of all the transformations which have acted on H. This aspect of the method makes it clear once again that the method of using the simple transformations of type $\mathbf{1} + R(K)$ is not inextricably linked with the particular perturbation rule (6.7) for estimating the transformation elements.

JOLLY. Programming notes

6.10 The form of $R(N, K)$

In the algebraic theory of the preceding sections the notation $R(N, K)$ was used for the elements of the column $R(K)$. In the program it is quite sufficient to use a *single column* of N elements throughout for $R(K)$. The particular index K intended is specified by means of the range of the summation indices in the matrix operations, and so changes as different eigenvalues are calculated. All the other elements of the matrices R and S are either zero or one; they are not stored, but have been included explicitly in the form of the basic assignment statements (6.5) and (6.6) of the algorithm.

6.11 The matrix indices

The row and column indices on the $N \times N$ matrix H naturally run from 1 to N. However, in JOLLY it is possible to select the upper and lower indices (U and L) which are to be used in the calculation, so that a square submatrix of H is used in the calculation.

There are two reasons for the introduction of this flexibility. First, it is needed in order to exploit the speedy calculation of eigenvalues using the deflation method explained in §6.7. Second, if H is the Hamiltonian matrix for a quantum mechanical problem, we traditionally wish to explore how the lower eigenvalues vary as the dimension of the matrix increases. This can be done by setting the lower index L equal to one and gradually increasing the upper index U used in the eigenvalue calculation, provided of course that the basis set is ordered in terms of increasing energy (or, more generally, that $H(N, N)$ is an increasing function of N).

6.12 The working copy, including λ

Since the matrix H is transformed and thus would be destroyed during the calculation, it must be stored and left unchanged. A working copy C of H is made at the start of each individual eigenvalue calculation; it is C which is transformed to reduced form by the transformations.

The equations (6.7) and (6.8) which give the elements of $R(K)$ are clearly most accurate when the off-diagonal elements of H are small, although the idea of the method is to keep on repeating the transformation until those off-diagonal elements (in the column K) are reduced to zero. Grosjean and Jolicard (1987) suggested that, like the traditional perturbation theory (used in RALLY), the algorithm used here has a limited region of convergence. To explore this point, we have included in JOLLY the possibility of multiplying the off-diagonal elements of H by a factor λ. For small λ the method will converge to give a result after only a few iterations, but as λ is increased it could be that beyond some critical λ value no apparent convergence will be obtained even after many iterations. It was partly to alleviate this possible difficulty that the attenuation equation (6.8) was introduced; we certainly do not find the problem to be as acute as Grosjean and Jolicard (1987) suspected, at least for the *eigenvalue* calculation using the deflation process.

6.13 The subroutine structure

The program is designed so that the computation of the elements of $R(K)$ is confined to one subroutine while the application of the matrix transformation *using* $R(K)$ is confined to another subroutine. This decoupling of the two operations makes it possible to invoke the transformation subroutine by passing to it an $R(K)$ which was *not* obtained from the perturbation prescription. It thus becomes possible to apply the kind of preliminary transformation mentioned in §6.9. If we already know an approximate eigencolumn to use in the transformation, then the preliminary transformation will take H (or C) almost into reduced form. The job can then be finished off quickly using the perturbative formulae.

6.14 JOLLY. Program analysis and program

Lines 10 to 70 set up the matrix array for H and for the working copy C, together with the columns R and X.

Lines 100 to 150 set up the copy matrix C with off-diagonal elements which are λ times the off-diagonal elements of H.

Lines 200 and 205 set the lower and upper indices (L and U) which select a submatrix of H for calculation, together with the index K which selects a particular eigenvalue. The index K actually selects a diagonal element H(K, K) as the initial eigenvalue estimate, with the assumption that the Kth eigenvalue arises from it under the action of the off-diagonal elements regarded as a perturbation.

Lines 210 to 245 perform the pretransformation on H using (if I = 1) for the principal column of R the column X as produced by the previous calculation. If I $\neq$ 1 the elements of R are set to zero. Line 240 sets the Kth element of R and X equal to one. It is intended, of course, that K shall be kept the same throughout if I is to be chosen as one, since the X from the previous calculation is then to be used to give a reasonable start to the current calculation. Most commonly, the current calculation differs from the previous one by having an increased value of λ. If I = 1 then line 245 ensures that the transformation subroutine is used once before the calculation of R(K) begins to use the first-order perturbation rule.

Lines 270 and 280 display the current estimate of the energy, that is the diagonal element C(K, K) of the transformed matrix.

Subroutine 300 to 370 works out the elements of the R column using the perturbation prescription. Line 320 as shown represents a departure from the theory of §6.4, since it uses the fixed H(K, K) instead of the variable C(K, K) in the denominator of the perturbation expression for R. This amounts to using a perturbative Rayleigh–Schrödinger formula instead of a Brillouin–Wigner one. In some cases the use of H(K, K) has been found empirically to work better, but C(K, K) can be inserted easily into line 320 if required. In line 340 also a departure from the theory is present, with a two in the denominator instead of a one. This change has also been found to improve convergence in some cases, but a one can easily be used if required. Line 350 adds the current R into X to build up the elements of the eigencolumn.

Subroutine 400 to 530 carries out the transformation $C \rightarrow S^{-1}CS$, with $S = \mathbf{1} + R$. The two stages of the transformation, lines 400 to 450 and 470 to 530, correspond exactly to the two stages of the algorithm as set out in §6.4.

Lines 600 to 630 print out the eigencolumn X when called by the manual command GO TO 600.

The version given here is controlled by manual commands, but can be made more automatic if desired. For example a line 265 can be inserted which checks whether |1 − E/C(K, K)| is less than (say) 2E − 8 and jumps to 600 if it is.

```
 5 REM  ***********************************
 6 REM  JOLLY
 7 REM  ***********************************
 8 REM  INPUT MATRIX
 9 REM  ***********************************
10 INPUT "DIMENSION";Q
20 DIM H(Q,Q): DIM C(Q,Q): DIM R(Q): DIM X(Q)
30 PRINT "INPUT MATRIX H"
40 FOR M=1 TO Q: FOR N=1 TO Q
50 PRINT M;N: INPUT Z
60 PRINT Z: LET H(M,N)=Z
70 NEXT N: NEXT M
97 REM  ***********************************
```

```
 98 REM   COPY H INTO C
 99 REM   *********************************
100 INPUT "LAMDA";LA
110 FOR M=1 TO Q: FOR N=1 TO Q
120 LET C(M,N)=H(M,N)*LA
130 NEXT N
140 LET C(M,M)=H(M,M)
150 NEXT M
199 REM   *********************************
200 INPUT "L AND U INDICES";L,U
205 INPUT "SELECTED INDEX";K
210 INPUT "1 FOR PRETRANS";I
215 FOR J=L TO U
220 LET S=X(J)*(I=1)
225 LET R(J)=S
230 LET X(J)=S
235 NEXT J
240 LET R(K)=1: LET X(K)=1
245 IF I=1 THEN GO TO 260
250 GO SUB 300
260 GO SUB 400
270 LET E=C(K,K): PRINT E
280 GO TO 250
297 REM   *********************************
298 REM   R COLUMN CALCULATION
299 REM   *********************************
300 FOR M=L TO U
310 IF M=K THEN GO TO 360
320 LET ET=H(K,K)
330 LET R=C(M,K)/(ET-C(M,M))
340 LET R(M)=R/(2+ABS R)
350 LET X(M)=X(M)+R(M)
360 NEXT M
370 RETURN
397 REM   *********************************
398 REM   RHR TRANSFORMATION
399 REM   *********************************
400 FOR M=L TO U: LET S=0
410   FOR N=L TO U
420 LET S=S+C(M,N)*R(N)
430   NEXT N
440 LET C(M,K)=S
450 NEXT M
460 REM   *********************************
470 FOR M=L TO U
480 IF M=K THEN GO TO 520
490   FOR N=L TO U
500   LET C(M,N)=C(M,N)-C(K,N)*R(M)
```

```
510     NEXT N
520 NEXT M
530 RETURN
597 REM  ***********************************
598 REM  EIGENCOLUMN
599 REM  ***********************************
600 FOR M=L TO U
610 PRINT X(M)
620 NEXT M
630 GO TO 200
631 REM  ***********************************
```

JOLLY. Specimen results

6.15 A 5 × 5 example

The first test example is a 5 × 5 matrix with the elements

```
H(J,K)=J*(J=K)+(J<>K)                                        (6.15)
```

This Boolean function formulation can be used to compute the elements without using the manual input, after a simple modification of line 50 which computes H(J, K) instead of asking for an input. By using varying λ values in line 100 the off-diagonal elements are all made equal to λ, while the diagonal elements have integer values from one to five which are λ independent. We describe the results of the calculation for three different values of λ. The choice ET = C(K, K) was used in line 320.

$\lambda = 1$. Using the denominator $2 + |R|$ in line 340 gives convergence for each K value, with GO TO 100 being used to choose another K after the previous eigenvalue has converged. If the denominator $(1 + |R|)$ is used, then only the largest eigenvalue (K = 5) converges. It should be noted, however, that *if* the calculation converges with the latter denominator (as it does for $\lambda = 1/2$) then it does so more rapidly, as would be expected from the mathematical theory.

$\lambda = 2$. Even with the value 2 in the denominator of line 340, the results for K = 1 and K = 4 are not obtainable, because of divergence or (at best) very slow convergence. The results for K = 2, 3 and 5 are obtained satisfactorily, with the greatest eigenvalue (K = 5) converging most rapidly. To obtain all five eigenvalues we first compute the K = 5 eigenvalue, then use GO TO 200 followed by L = 1, U = 4, K = 4. This gives the next eigenvalue; we then use GO TO 200 and L = 1, U = 3, K = 3 and so on, each successive eigenvalue being obtained more quickly. The last step, L = U = K = 1 produces the last eigenvalue immediately. Since the program has not been designed to perform eigencolumn deflation (see §6.6) the eigencolumns for K = 1 and 4 are not obtained.

$\lambda = 5$. Here only the largest (K = 5) eigenvalue is obtainable together with its eigencolumn, but eigenvalue deflation then gives the other four eigenvalues quickly.

Table 6.1 shows the eigenvalues for $\lambda = 1$, 2 and 5.

A check on the eigenvalues is obtained by noting that they must add up to give the diagonal sum of the matrix, in this case 15. For the special test matrix used here, with all off-diagonal elements equal, Sparks (1970) has proved a theorem which gives bounds on the eigenvalues; the first four eigenvalues fall within bands of unit width starting at $1 - \lambda$.

Table 6.1 Eigenvalues for the test 5 × 5 matrix.

K	$\lambda = 1$	2	5
1	0.277 695 8	−0.686 757 0	−3.662 207 5
2	1.356 631 9	0.403 787 0	−2.565 321 0
3	2.434 736 7	1.488 954 6	−1.477 933 0
4	3.540 394 4	2.595 211 2	−0.374 461 7
5	7.390 541 2	11.198 804	23.079 923

6.16 Matrices with equal diagonal elements

The perturbation formula for the elements of the column R will give a divergence for the case C(K, K) = C(M, M). JOLLY is best suited to matrices which have different elements H(N, N) and small off-diagonal elements. To test the method on a matrix which strongly violates these conditions we applied it to a 3 × 3 matrix with all three diagonal elements zero and with off-diagonal elements H(J, K) = J + K. The use of *any* K value (1, 2 or 3) in the iterative computation produces the largest eigenvalue 8.055 810 4. Eigenvalue deflation using GO TO 200 gives the other two eigenvalues −5.180 267 8 and −2.875 542 6. The sum of the eigenvalues is zero, as required by the diagonal sum rule. To obtain the preceding results it was necessary to modify the program as follows. In the subroutine which calculates the R column the following lines were used.

```
320 LET ET=C(K,K)
325 LET D=ET-C(M,M)                                   (6.16)
326 IF D=0 THEN LET D=IE-8
330 LET R=C(M,K)/D
```

This new formulation avoids the use of exactly zero divisors; even a very small divisor produces a bounded R(M) value in line 340. The resulting program gives all three eigenvalues for our 3×3 test case, but only one eigenvector, namely that for the largest eigenvalue.

6.17 Using Aitken extrapolation

The successive values of C(K, K) converge towards the required eigenvalue E(K) with errors which vary in a roughly geometric progression. The use of Aitken extrapolation on the sequence of E estimates usually gives a more quickly convergent sequence of estimates, although the accurate eigencolumn only emerges after the primary sequence has converged. To obtain accurate values from the Aitken extrapolation the more effective form of the extrapolation formula must be used, that is sets of three successive values A, B, E must be combined using the expression

$$\mathrm{EP} = \mathrm{A} - (\mathrm{A} - \mathrm{B})^2(\mathrm{A} + \mathrm{E} - 2\mathrm{B})^{-1}. \tag{6.17}$$

This can be used in line 275, with the assignments A=B and B=E in line 265. The print commands can be modified to show E and EP (E projected) side by side on the screen; this displays clearly how EP often has several more converged digits than E. To avoid a program halt because of unspecified variables it is necessary to have an earlier line which assigns A, B and E some trivial starting values, which means that the first three EP values are not valid estimates of the converged eigenvalue.

Several *ad hoc* tricks can be used to obtain convergence in some cases. For example the use of a denominator $(3 + |\mathrm{R}|)$ in line 340 and of ET=H(K, K) in line 320 are sufficient to produce the $\mathrm{K} = 1$, 2 and 3 eigencolumns (from the standard process with N=5) at $\lambda = 2$ for our first 5×5 test matrix. It seems possible that a more complicated rule for choosing the R(K) elements could be formulated to incorporate these adjustments, which were discovered by interactive computational experiment. The method of JOLLY, being comparatively new, is still being investigated and could well be capable of improvements which extend the region of convergence for eigencolumn calculation without incorporating the full-scale eigencolumn deflation approach outlined in §6.6.

FOLDER. Mathematical theory

6.18 Introduction

The folding algorithm used in FOLDER has its origins in the ideas of perturbation theory (Feenberg 1948, Feshbach 1948, Löwdin 1951) and the relevant formulae are often presented in a form in which they resemble traditional perturbation formulae. However, an investigation of the underlying mathematical principles reveals that the techniques involved are *not* limited by any considerations about the smallness of some perturbation parameter; in fact, they arise naturally out of a study of methods for solving systems of linear equations. Some of the mathematical discussion used in setting up FOLDER is based on a synthesis of ideas which seems to the author to be original.

6.19 The perturbation approach

The matrix eigenvalue problem

$$Hx = Ex \tag{6.18}$$

where H is an $N \times N$ matrix, can be written out in a partitioned form, with the Nth row and column being treated separately from the rest of the matrix. This is often shown pictorially by splitting the matrix up into sections. However, we can derive the required algorithm better by performing the partitioning algebraically. This simply involves isolating the index N when writing down the eigenvalue equations. For $J = 1$ to $N - 1$ we have

$$\sum_{1}^{N-1} H(J, K)X(K) + H(J, N)X(N) = EX(J) \tag{6.19}$$

while for $J = N$ we have

$$\sum_{1}^{N-1} H(N, K)X(K) + H(N, N)X(N) = EX(N). \tag{6.20}$$

Equation (6.20) provides an expression for $X(N)$ which can be substituted in (6.19) to remove $X(N)$. Equation (6.19) is thus converted into an $(N - 1) \times (N - 1)$ matrix eigenvalue problem

$$\sum_{1}^{N-1} H(J, K)X(K) = EX(K) \tag{6.21}$$

where the elements of H have been modified by the folding rule assignment statement

$$H(J, K) := H(J, K) + H(J, N)H(N, K)D^{-1} \tag{6.22}$$

with the denominator

$$D = E - H(N, N). \tag{6.23}$$

The result (6.22) is highly reminiscent of the formulae of second-order perturbation theory. As an eigenvalue equation, however, (6.21) is an *implicit* equation, since E must be varied until it equals an eigenvalue of a matrix which itself contains E. The result is a self-consistent eigenvalue calculation (Williams and Weaire 1976), but it is still a difficult matrix problem. The obvious way to avoid this is to *keep on folding*, until the matrix is reduced to size 1×1. If the final value of that $(1, 1)$ element equals E, then E is an eigenvalue of the original $N \times N$ matrix. If it is not, then the input E value is varied until the difference $H(1,1) - E$ is zero. For an $N \times N$ matrix there will be N roots to the scalar equation $H(1,1) - E = 0$, giving the N eigenvalues of the original $N \times N$ matrix. This approach via total folding of the matrix (Killingbeck 1985a) has the advantage of transforming the matrix eigenvalue problem into a root-finding problem. It also involves a simple algorithm, since the folding operation only requires the repeated use of equations (6.22) and (6.23) to modify the matrix elements.

6.20 An explicit example

Feenberg (1948) and Feshbach (1948) proposed a version of perturbation theory which includes special rules for calculating higher-order terms in the energy perturbation series. These rules ensure that for a finite matrix the series terminates and produces an implicit equation from which the matrix eigenvalues are calculable in principle. Inspection of their results indicates that the folding process of the last section produces the same implicit equation, without the use of any rules other than the simple repetitive folding rule.

The folding process is usually carried out numerically, but can also be performed algebraically. For a test matrix of the form

$$H = \begin{pmatrix} 1 & \lambda & \lambda \\ \lambda & 2 & \lambda \\ \lambda & \lambda & 3 \end{pmatrix} \qquad (6.24)$$

the folding process leads to the implicit equation

$$E = 1 + \lambda^2[2E + 2\lambda - 5][(E - 2)(E - 3) - \lambda^2]^{-1}. \qquad (6.25)$$

Multiplying this equation out gives a cubic equation, which at $\lambda = 1$ has the approximate roots 0.325, 1.460 and 4.214. For $\lambda \to 0$ we obtain $E = 1 - 1.5\lambda^2 + \ldots$, the usual perturbation series for the *particular* eigenvalue arising from the unperturbed eigenvalue 1. However, the equation (6.25) produces *all three* eigenvalues if solved fully, and does so for any value of the perturbation parameter λ. This example illustrates the fact that, despite the appearance of equation (6.22), the folding approach is not subject to the usual convergence difficulties which are associated with perturbation methods.

If the folding process is done downwards on to the (3, 3) element, instead of upwards on to the (1, 1) element, then the resulting implicit equation is

$$E = 3 + \lambda^2[2E + 2\lambda - 3][(E - 1)(E - 2) - \lambda^2]^{-1}. \qquad (6.26)$$

For $\lambda \rightarrow 0$ this produces the perturbation series for the eigenvalue with unperturbed value 3, but for arbitrary λ it gives the same cubic equation as that arising from (6.25).

Equations (6.25) and (6.26) reveal a feature of the folding method which is difficult to understand from a perturbation theory viewpoint, but which is understandable using the alternative approach of the next section. The function on the right of (6.25) has two singularities, one at each of the zeros of the quadratic expression $(E - 2)(E - 3) - \lambda^2$. Inspection of the matrix H reveals that these zeros fall at E values which are exactly the two eigenvalues of the 2×2 submatrix of H obtained by deleting the first row and column of H. The function in (6.25) thus contains information about those eigenvalues as well as those of H. This apparent bonus has a drawback, however. If both the submatrix and H have some eigenvalues which are close together, the root-finding algorithm has the tricky job of finding a root which is very close to a singularity. Peters and Wilkinson (1969) commented on this kind of difficulty in a paper dealing with the generalized eigenvalue problem for symmetric matrices. When H is the Hamiltonian matrix for a quantum mechanical problem, it is usual to order the diagonal elements so that they increase monotonically. In that case folding up on to the (1, 1) element, in accord with the equations of §6.19, somewhat alleviates the difficulty. Folding down on to the (N, N) element *would* be likely to give problems, since the whole point of the quantum mechanical calculation is to increase the dimension until the $N \times N$ and $(N - 1) \times (N - 1)$ matrices *do* have almost identical eigenvalues! The traditional methods which use special recurrence relations in calculating matrix eigenvalues, such as the recurrence relations for tridiagonal or quindiagonal matrices (Barth *et al* 1967, Evans 1973, Makarewicz 1989) can be shown to be special cases of the folding method with the folding being done on to the (N, N) element. When the matrix H is of banded form, each folding transformation (6.22) only modifies a small number of elements; thus, for a pentadiagonal matrix only four elements of H have to be changed in each folding operation. The folding transformation (6.22) thus appears to be the most simple way of generalizing the various special recurrence relation methods which have been used previously.

6.21 An alternative approach

To clarify some of the features of the folding method it is best to start from the problem of solving a set of N simultaneous linear equations

$$Mx = y \tag{6.27}$$

with the column y given. Making the same partitioning assumption which was made for the calculations of §6.19 we obtain the following results:

For $J = 1$ to N − 1:

$$\sum_{1}^{N-1} M(J, K)X(K) + M(J, N)X(N) = Y(J). \tag{6.28}$$

For $J = N$:

$$\sum_{1}^{N-1} M(N, K)X(K) + M(N, N)X(N) = Y(N). \tag{6.29}$$

Finding $X(N)$ from (6.29) and substituting in (6.28) gives the reduced system of $N - 1$ equations

$$Mx = y$$

with the elements modified according to the assignment statements

$$M(J, K) := M(J, K) - M(J, N)M(N, K)D^{-1} \tag{6.30}$$

and

$$Y(J) := Y(J) - Y(N)D^{-1} \tag{6.31}$$

where

$$D = M(N, N). \tag{6.32}$$

If the folding down of M and y is carried through to the 1×1 stage, then the element $X(1)$ follows at once from that final 1×1 equation. More interestingly, the rest of the X column can be reconstructed, using *only* information which is stored in the transformed H array. To see this, we require the equation which gives $X(N)$, that is the rearranged version of equation (6.29). This is (with J replacing N) the assignment

$$X(J) := \left(Y(J) - \sum_{1}^{J-1} M(J, K)X(K)\right) \Big/ M(J, J). \tag{6.33}$$

By checking through simple 2×2 and 3×3 examples the reader can establish that (6.33) does indeed correctly give the elements $X(J)$ for $J = 2$ to N, starting from $X(1)$ as given by the final 1×1 equation which terminates the folding process. The elements $M(J, K)$ needed are just those already modified and retained during the folding process.

The interpretation of the folding algorithm as a method for solving a system of linear equations throws new light on the eigenvalue calculation discussed in the preceding sections. If we suppose that all the $Y(J)$ are zero except $Y(1)$, which is one, then it follows that the $M(1,1)$ element obtained at the end of the folding process obeys the equation

$$M(1, 1)X(1) = 1. \tag{6.34}$$

Thus the folded $M(1,1)$ is the reciprocal of the $(1,1)$ element of the inverse matrix M^{-1}. If the matrix M is singular (i.e. has a zero determinant) then M^{-1} acquires formally infinite elements, so that $M(1,1)$ becomes zero. Thus the result $M(1,1) = 0$ *at the end of the folding process* implies that M has a zero determinant.

If we wish to solve the generalized eigenvalue problem

$$Hx = ESx \tag{6.35}$$

where H and S are both $N \times N$ matrices, then we make the choice $M = H - \lambda S$, with λ some trial value for E. The folded $M(1,1)$ then becomes a function of λ and the eigenvalues of the problem are the λ values for which $M(1,1)$ is zero. If $M(2,2)$ is zero, it is clear that $M(1,1)$ will become formally infinite; this is just the case discussed in §6.20 in which the $(N-1) \times (N-1)$ submatrix has an eigenvalue. In fact, if we recall the textbook definition of the inverse matrix M^{-1} in terms of the adjoint matrix, we can see that the $M(1,1)$ element obtained after folding is the ratio of the determinants of M and the $(N-1) \times (N-1)$ submatrix. More generally, the diagonal elements $M(J,J)$ derived during the folding process are the ratios of principal minors of the original matrix M. This fact means that the appearance of a zero in any of these intermediate diagonal elements could cause disaster. However, it also provides a commonsense way of avoiding the difficulty and of removing the singularities which might appear in the quantity $M(1,1)$. We use two combined devices. First, we *multiply* together the $M(J,J)$ elements obtained during the folding process. If we include the original (unmodified) $M(N,N)$ element we obtain the product

$$\frac{|N|}{|N-1|} \times \frac{|N-1|}{|N-2|} \times \ldots \times \frac{|2|}{|1|} \times |1| \tag{6.36}$$

where $|J|$ denotes the determinant of the principal $J \times J$ submatrix of M. The product (6.36) is just the determinant of the matrix $M = H - \lambda S$, and we require it to be zero in order to locate the eigenvalues of the generalized eigenvalue problem (6.35). The second device is used if, during the folding, one of the intermediate divisors D used in (6.32) happens to be zero (or *very* small); we then replace it by $1E-8$. This allows the calculation to proceed and is simply equivalent to adding 1E − 8 to one of the diagonal elements of H, albeit only for the particular trial E value which is causing the difficulty by being an eigenvalue of one of the submatrices.

The two devices explained above are sufficient to render the folding method very effective for calculating the eigenvalues of the kind of quantum mechanical matrix which is considered in this book. It is clear from the discussion that the method will work just as well for the generalized eigenvalue problem (6.35) as for the ordinary eigenvalue problem in which S is the unit matrix. As the folding formula (6.30) indicates, the method can be applied to real non-symmetric matrices, yielding whatever real eigenvalues there are to be found. We should note that the method finally adopted here actually amounts to the use of the folding algorithm to work out the determinant of $H - \lambda S$. The evaluation of this determinant is a widely used method of calculating eigenvalues; the novelty here is in the use of the very simple repetitive folding algorithm which generalizes several of the recursive algorithms used previously for special types of matrix. We have presented the theory here in a manner which indicates its origin in work on perturbation theory and on linear equation theory. It is perhaps worth noting that the theory could also be based directly on the theory of determinants by starting from the remarkable identity quoted by Shishov (1961) for the determinant of a partitioned matrix:

$$\begin{vmatrix} A & B \\ C & D \end{vmatrix} = |A - BD^{-1}C|\,|D|. \tag{6.37}$$

6.22 Calculating eigenvectors

As the folding process proceeds, the elements of H are progressively modified but retain their position in the H array. After the run in which E is found (i.e. in which the determinant is as close to zero as can be obtained with the computer being used) the eigencolumn can be constructed by setting $X(1) = 1$ and then using the assignment statement (6.33) with all the $Y(J)$ set to zero. This gives (for $J = 2$ to N)

$$X(J) := -\sum_{1}^{J-1} M(J, K)X(K)/M(J, J). \tag{6.38}$$

To understand the process the reader may think of it as a form of inverse iteration (Wilkinson 1965) in which the matrix $(H - ES)^{-1}$ acts on the column $(1, 0, 0, \ldots)^{\mathrm{T}}$ to produce an eigencolumn, which is here normalized to have $X(1) = 1$. To make formal sense of traditional inverse iteration the E value used in $(H - ES)^{-1}$ should differ very slightly from the eigenvalue. However, the prescription (6.38) works correctly; as the equations show, all the $X(J)$ are proportional to $X(1)$ and so we can set it equal to one for convenience. Even though the determinant $|H - ES|$ is zero, the matrix $H - ES = M$ gives well behaved results when used in (6.38).

6.23 Solving linear equations

The theory of §6.21 shows that the folding algorithm is a multi-purpose one. It can find matrix eigenvalues and eigencolumns, calculate determinants and solve systems of linear equations. This means that the main folding subroutine used in FOLDER can be used in combination with other modules to perform a variety of tasks. For example, the inverse H^{-1} can be calculated by solving the equations $Hx = y$, with y being set equal in turn to the unit columns $e(J)$ which have a one in the position J and zeros elsewhere. The author has tested a program LINFOLD which performs this task using the folding transformation, but has found the program MATIN of this book to be more speedy. Rao (1989) described an approach to the solution of systems of simultaneous linear equations which uses what is essentially the folding transformation. The method of Aasen and Romberg (1965) for eigenvalue calculation has some similarity to that used in FOLDER, but produces a function $F(E)$ with tan-type singularities.

FOLDER. Programming notes

6.24 Array requirements

The matrix H of $N \times N$ type must be stored and kept intact throughout the computation. A working copy C of H is made before each folding operation, and it is C which is modified during the folding process. A column X of N elements is needed to contain the eigencolumn. X can be reused for the various different eigencolumns or can be set up as an $N \times N$ array if it is necessary to store all the eigencolumns (or to store the inverse of H, using the approach mentioned in §6.23).

6.25 The modular structure

The folding algorithm will be implemented within a subroutine which takes the input λ (actually called E in the program) and returns the value of the determinant of $H - ES$ or of the folded $M(1,1)$ element if required. Associated with the subroutine are the modules which are common to several programs. These are: the matrix input routine, the ROOTSCAN routine and the root-finder routine. New modules which could be added include those which perform the following tasks: computation of the eigencolumn, solution of the equation system $Hx = y$, and computation of the inverse matrix H^{-1}. All of these new modules could use the central folding subroutine, but only the eigencolumn module is included in the FOLDER listing shown here.

6.26 Index permutation

The mathematical theory set out so far has assumed that folding is carried out on to the (1, 1) element. It is clear how to fold down to the (N, N) element instead, but how would we fold on to the (3, 3) element, say, in a 6×6 matrix? The outer layers could be 'peeled off' from above and below in several different orders. To allow the effect of folding on to the diagonal element $H(I, I)$ to be explored we can use a standard prescription: we simply exchange the roles of index 1 and index J in the copy matrix C before folding commences. The program can then retain its fixed form, always folding on to the (1, 1) element. The reader may check by constructing examples or by looking at the relevant transformation matrices that the roles of indices 1 and I are reversed by performing the following two steps (each of which needs a temporary store element T to hold an element): first interchange row 1 and row I of C to get C', then interchange column 1 and column I of C' to get the required result.

6.27 The use of submatrices

To make it possible to study submatrices of the $N \times N$ matrix H, the operator is allowed to choose the lower and upper indices (L and U) to be used in the calculation. When L is not equal to one, the folding is done on to the (L, L) element, and so the permutation discussed in §6.26 becomes $L \leftrightarrow I$ instead of $1 \leftrightarrow I$. Reasons for studying submatrices of H were explained in connection with the program JOLLY. In the case of FOLDER it is sometimes of interest to investigate the cases (mentioned earlier) in which a principal submatrix of H has an eigenvalue close to an eigenvalue of H.

6.28 FOLDER. Program analysis and program

Lines 10 and 20 set up the N × N matrix arrays for H and S and for the working copy C to be used during the folding operation. The column X is set up to store the elements of the particular eigencolumn being calculated.

Lines 30 to 50 are the input lines for the matrix H. In line 45 the matrix S is set equal to the unit matrix by using a Boolean function. This means that the eigenvalue equation $Hx = ESx$ has been provisionally set up with $S = \mathbf{1}$, which is the case most commonly encountered.

Lines 60 to 90 are the input lines for S, but only need to be used if S is *not* the unit matrix. The number I is given by the operator (in line 60) to indicate whether the S input can be skipped.

Lines 200 and 210 allow a choice of the lower (L) and upper (U) indices to be used. This allows calculations to be performed on submatrices, so that the effect on the eigenvalues of varying the matrix dimension can be investigated. Line 210 sets the particular diagonal element (I, I) on to which the folding is to be done.

Line 240 sets the initial eigenvalue guess E and the step length DE to be used by the ROOTSCAN module.

Lines 300 to 360 carry out the scannning process until the function of E shows a sign change between E = E1 and E = E + DE = E2.

Lines 400 to 470 constitute the secant method root finder, which produces a refined root in the region between E1 and E2. Line 450 produces a jump to the eigencolumn calculation when E has converged, and also sets up the value E = E2 to continue the root-scanning process later. Line 455 guards against the possibility of a very small local gradient of the function producing a shift which jumps right out of the interval of length DE in which the desired root must be located.

Lines 500 to 590 work out the eigencolumn, with X(L) = 1, using the algorithm explained in §6.22. Line 570 interchanges elements X(L) and X(I) if the folding was done on to an element with I ≠ L. The single line loop in line 580 can, of course, be expanded to three lines if desired. Line 590 returns the calculation to ROOTSCAN at E = E2.

Lines 1000 to 1020 make the working copy C of H – ES to be used in the folding operation.

Line 1025 goes to subroutine 2000 if the interchange L ↔ I is needed.

Lines 1030 to 1100 perform the folding operation. Note lines 1035 and 1045, which avoid the presence of a zero divisor (see §6.21). Line 1055 sets up the factor T which is constant throughout the following K loop, and thus avoids the redundant repeated calculation of it inside the loop. The subroutine returns the value of the determinant |H – ES| to the root-finding modules.

Lines 2000 to 2060 interchange the roles of the indices L and I in the copy matrix C, as explained in the programming notes.

```
  7 REM    ********************************
  8 REM  FOLDER
  9 REM    ********************************
 10 INPUT "DIMENSION";N
 20   DIM H(N,N): DIM S(N,N): DIM C(N,N): DIM X(N)
 21 REM    ********************************
 30 PRINT "INPUT MATRIX H"
 31 REM    ********************************
 35 FOR J=1 TO N: FOR K=1 TO N
 40 PRINT J;K: INPUT X: PRINT X
 45 LET H(J,K)=X: LET S(J,K)=(J=K)
 50 NEXT K: NEXT J
 51 REM    ********************************
 60 PRINT "INPUT 1 IF S IS UNIT MATRIX": INPUT P
 65 IF P=1 THEN GO TO 200
 69 REM    ********************************
 70 PRINT "INPUT S MATRIX"
 71 REM    ********************************
 75 FOR J=1 TO N: FOR K=1 TO N
 80 PRINT J,K: INPUT X: PRINT X
 85 LET S(J,K)=X
 90 NEXT K: NEXT J
 91 REM    ********************************
200 INPUT "INDICES L,U";L,U
210 INPUT "FOLD INDEX";I
240 INPUT "E0,DE";E0,DE
300 LET E=E0
301 REM    ********************************
302 REM  ROOTSCAN
303 REM    ********************************
310 GO SUB 1000: PRINT E,F
320 LET F1=F: LET E1=E
330 LET E=E+DE
340 GO SUB 1000: PRINT E,F
350 LET F2=F: LET E2=E: LET R=F2/F1
360 IF R>0 THEN GO TO 320
361 REM    ********************************
398 REM  SECANT
399 REM    ********************************
400 LET E=E2-DE+DE/(1-R)
405 LET FS=F2: LET ES=E2
410 GO SUB 1000: PRINT E,F
430 LET GR=(F-FS)/(E-ES)
440 LET ES=E: LET FS=F
445 LET SH=F/GR
450 IF ABS (SH/E)<2E-8 THEN LET E=E2: GO TO 500
455 IF ABS SH>ABS (DE/4) THEN LET SH=ABS (DE/4)*SGN SH
460 LET E=E-SH
```

```
470 GO TO 410
497 REM  *********************************
498 REM  COLUMN CALC.
499 REM  *********************************
500 LET X(L)=1: PRINT "COLUMN"
510 FOR J=L TO U-1: LET S=0
520   FOR K=L TO J
530 LET S=S-C(J+1,K)*X(K)
540   NEXT K
550 LET X(J+1)=S/C(J+1,J+1)
560 NEXT J
570 LET T=X(I): LET X(I)=X(L): LET X(L)=T
580 FOR J=L TO U: PRINT X(J): NEXT J
590 GO TO 310
591 REM  *********************************
998 REM  FOLD TRANS.
999 REM  *********************************
1000 FOR J=L TO U: FOR K=L TO U
1010 LET C(J,K)=H(J,K)-E*S(J,K)
1020 NEXT K: NEXT J
1021 REM  *********************************
1025 GO SUB 2000
1030 LET DET=C(U,U)
1035 IF DET=0 THEN LET DET=1E-8
1039 REM  *********************************
1040 FOR M=U TO L+1 STEP -1
1045 IF C(M,M)=0 THEN LET C(M,M)=1E-8
1050   FOR J=L TO M-1
1055 LET T=C(J,M)/C(M,M)
1060   FOR K=L TO M-1
1065 LET C(J,K)=C(J,K)-T*C(M,K)
1070   NEXT K: NEXT J
1075 LET DET=DET*C(M-1,M-1)
1080 NEXT M
1081 REM  *********************************
1090 LET F=DET
1100 RETURN
1101 REM  *********************************
1998 REM  INDEX PERMUTATION
1999 REM  *********************************
2000 FOR J=L TO U
2010 LET T=C(J,I): LET C(J,I)=C(J,L): LET C(J,L)=T
2020 NEXT J
2030 FOR J=L TO U
2040 LET T=C(I,J): LET C(I,J)=C(L,J): LET C(L,J)=T
2050 NEXT J
2060 RETURN
2061 REM  *********************************
```

FOLDER. Specimen results

6.29 Simple test example

For the 5 × 5 and 3 × 3 test matrices which were used to test JOLLY, the program FOLDER quickly gives the eigenvalues and eigencolumns when a DE value of 1/2 is used, with E0 sufficiently low to find the first eigenvalue. In general, of course, DE must be smaller than the eigenvalue separation, but for bound state matrix problems in quantum mechanics the eigenvalues often show a reasonable separation. For the 3 × 3 matrix with zero diagonal elements and the off-diagonal elements H(J, K) = J + K, FOLDER gives the results shown in table 6.2. The results of table 6.2 were obtained using the fixed folding index I = 1 which is why X(1) = 1 in all three cases. The three eigencolumns are orthogonal (to within rounding error), as is required for the case when H is symmetric.

Table 6.2 Eigenvalues and eigencolumn for test matrix.

E	−2.875 542 6	−5.180 267 8	8.055 810 4
X(1)	1	1	1
X(2)	−0.679 768 04	2.429 849 3	1.107 061 9
X(3)	−0.209 059 61	−3.117 454	1.183 656 2

6.30 A complex matrix example

Although we suppose throughout this book that only real arithmetic is used in the computations, it turns out that complex (and particularly Hermitian) matrices *can* be handled by FOLDER. The secret (once again!) lies in getting the algebra right before looking at the program. If we suppose that both the matrix and the eigencolumn can be complex, we can write the matrix eigenvalue problem in the shorthand symbolic form

$$(R + \mathrm{i}I)(X + \mathrm{i}Y) = E(X + \mathrm{i}Y) \tag{6.39}$$

where R and I are real $N \times N$ matrices and X and Y are real columns of N elements. On separating out the real and imaginary parts of the equation (6.39) we obtain a pair of simultaneous equations which can be written as a matrix eigenvalue problem for a matrix of dimension $2N \times 2N$:

$$\begin{pmatrix} R & -I \\ I & R \end{pmatrix}\begin{pmatrix} X \\ Y \end{pmatrix} = E\begin{pmatrix} X \\ Y \end{pmatrix}. \tag{6.40}$$

If the eigenvalue E is real, then this problem can be handled by FOLDER, although some further thought about the root-finding module is necessary.

6.31 Finding double roots

Equation (6.40) would be expected to have $2N$ eigenvalues, whereas the problem (6.39) will only have N eigenvalues. However, the eigenvalues of (6.40) are two-fold degenerate, so that only N distinct eigenvalues emerge. This causes difficulties for the root-finding module in FOLDER, since at a double root of a function $F(E)$ there is *no sign change* of F as E increases through the eigenvalue. Near the root the graph of the function shows a parabolic form. To find such a zero a two-stage process can be used. The simple 2×2 Hermitian matrix

$$H = \begin{pmatrix} 1 & 1+\mathrm{i} \\ 1-\mathrm{i} & 3 \end{pmatrix} \tag{6.41}$$

will provide an illustrative example. It leads to the 4×4 symmetric matrix eigenvalue problem for the matrix

$$\begin{pmatrix} 1 & 1 & 0 & -1 \\ 1 & 3 & 1 & 0 \\ 0 & 1 & 1 & 1 \\ -1 & 0 & 1 & 3 \end{pmatrix}. \tag{6.42}$$

Setting E0 = 0, DE = 0.1 in FOLDER reveals that the computed function F(E) is positive for all E but has minima, at which F(E) is very small, at E values of roughly 0.3 and 3.7. That is a good first result, since from the diagonal sum of H we can see that we are looking for two eigenvalues which add up to give a value 4.

Having found the approximate eigenvalues, we now need some method which is suitable for locating each particular eigenvalue by searching in a specified small region. The module ZIGZAG of chapter 2 is suitable, or we can modify the rootscan module in FOLDER. This simply requires the temporary insertion of the following two lines in the program:

```
355 IF R>1 THEN LET DE=-DE/4
357 IF ABS(E1/E2-1)<2E-8 THEN GO TO 500
```
(6.43)

These modifications make the program search backwards and forwards, homing in on the eigenvalue by using smaller and smaller search intervals. This modified method *also* works for single roots. To apply the method to the example being used here, appropriate starting parameters are E0 = 0.2, DE = 0.05 and E0 = 3.6, DE = 0.05. The modified program then produces the eigenvalue–eigenvector pairs

$$\mathrm{E} = 0.267\,949\,20$$

$$(1,\ -0.367\,924\,53,\ 0.005\,188\,502\,3,\ 0.364\,126\,28)^{\mathrm{T}}$$

and

$$\mathrm{E} = 3.732\,050\,9$$

$$(1,\ 1.367\,727\,8,\ 0.001\,246\,333\,7,\ -1.364\,322\,7)^{\mathrm{T}}.$$

These results can be interpreted for the original 2×2 matrix by recalling how the real and imaginary parts of the eigencolumns are defined in equation (6.39).

6.32 An iterative method

It has already been pointed out that the construction of the eigencolumn is essentially equivalent to solving the equation system

$$(H - ES)x = y \tag{6.44}$$

with $y = (Y, 0, 0 \ldots 0)^{\mathrm{T}}$, if we consider the case $I = 1$. Here Y is some number which we do not need to quote, since we always scale the *solution* up or down to make the element $X(1)$ equal to 1. The program module starting at line 500 will solve (6.44) for *any* given E, whether or not that E is an eigenvalue. This makes possible several kinds of iterative process. These are based on the idea that the solution x of (6.44), although it contains admixtures of several eigencolumns of H, will have a dominant contribution from one particular eigencolumn if E is close to the eigenvalue for that eigencolumn. The method of Rayleigh iteration proceeds by using the solution $x(E)$ of (6.44) to give a Rayleigh quotient $\langle H \rangle$ defined by

$$(x^{\mathrm{T}}x)\langle H \rangle = x^{\mathrm{T}}Hx \tag{6.45}$$

with $\langle H \rangle$ being taken as the new E value for use in (6.44). This method can be used with FOLDER, but a more simple approach is to work out the (1, 1) component of the matrix product HX, where X is the column produced by the module starting at the line 500. Since $X(1) = 1$, the evaluated number is an estimate of an eigenvalue, and will coincide with the input E when that input E is an eigenvalue.

A method based on iterations of the type described here does not need a full root-finding module. In principle, given E0, it might be expected to find its own way to the eigenvalue nearest to E0, producing the eigenvector at the same time. This idea can be tried by inserting the following temporary modifications in FOLDER:

```
315 GO TO 500
575 GO TO 582
582 LET S=0: FOR J=L TO U
584 LET S=S+H(I,J)*X(J)                                    (6.46)
586 NEXT J
587 LET RP=1/2
588 LET E=RP*S/X(I)+(1-RP)*E
```

Using this modification for the 5×5 test matrix with $H(J, K) = (J = K) + J*(J = K)$ gives the lowest or greatest eigenvalue, depending on whether a low or high E0 value is used. If the relaxation parameter RP is made smaller (typically 0.1) this tendency to move off to the extremal eigenvalue is damped; intermediate eigenvalues can then be picked out if E0 is sufficiently close to them. For the case of a symmetric matrix the use of the full Rayleigh quotient, although computationally more expensive than the approach tried here, would give even better stability in looking for intermediate eigenvalues. The Rayleigh quotient has the variational property of giving an eigenvalue error of order ε^2 if the approximate eigencolumn x has an error of order ε. To convert FOLDER permanently into a Rayleigh iteration program would best be done by writing a new subroutine to form the Rayleigh quotient, with most of the root-finding modules being discarded.

6.33 Band matrices

The program FOLDER as presented is applicable to real matrices which have non-zero elements in every position and have no symmetry properties. If the matrix H is symmetric, as is common in quantum or classical mechanics, it is not necessary to work out the folding matrix for elements below the diagonal, since the folding process preserves the symmetry of the matrix. To allow for this, the following simple modifications suffice in the folding routine:

Line 1055. C(M, J) replaces C(J, M).
Line 1060. J replaces M − 1.

Clearly, the effect of allowing for symmetry in this way is to roughly halve the computing time for each folding transformation, which represents a worthwhile saving.

If the matrix H is of banded type, so that for some small integer B we have

$$H(J, K) = 0 \qquad \text{if } |J - K| > B \tag{6.47}$$

then it follows that the number of elements modified at each step of the folding process is much reduced. Only B^2 elements are modified at each step, or $B(B+1)/2$ elements if H is symmetric. To fold down the matrix from $N \times N$ to 1×1 thus requires the modification of a total number of matrix elements which varies roughly as NB^2. This number grows linearly with N, whereas without banding the number of modifications grows roughly as $N^3/3$. To allow for the banded nature of H we can either modify the range of the indices in the relevant loops or test for a zero element and skip over the modification step if the test is obeyed. The former approach is more efficient for systematically banded matrices, although the latter can deal with cases in which zeros are scattered throughout an otherwise non-zero matrix. For the case of a tridiagonal matrix, which has $B = 1$, only one element (the diagonal one) is modified at each folding step. For the pentadiagonal case, with $B = 2$, only four elements are modified at each step. For a symmetric band matrix the restrictions on the sums over J and L can be accomplished, for example, by computing the limits before each loop. Thus, the lines

```
1047 LET JL=M-1-B
1048 LET JL=JL*(JL>L)+L*(JL<L)                    (6.48)
1050 FOR J=JL TO M-1
```

will set the correct J limit, JL, while the lines

```
1057 LET KL=J-B
1058 LET KL=KL*(KL>L)+L*(KL<L)                    (6.49)
```

will set the correct K limit, KL. The complicated form of the expressions arises because we must allow for the 'edge effect' when the bandwidth extends outside the matrix.

HITTER. Mathematical theory

6.34 The self-consistency problem

If the $N \times N$ matrix eigenvalue problem

$$Hx = Ex \tag{6.50}$$

is written in the form

$$\sum_{K=1}^{N} H(J, K)X(K) = EX(J) \tag{6.51}$$

then we can impose the specific requirement $X(M) = 1$ on an eigencolumn without loss of generality. If the off-diagonal elements of H are regarded as a perturbation, the choice $X(M) = 1$ is the natural one to make when studying the eigenvalue which arises from the unperturbed eigenvalue $H(M, M)$ as the off-diagonal elements are gradually increased. This perturbative approach is discussed in connection with the program RALLY in a later chapter. Having fixed $X(M)$, we can isolate the equation with $J = M$ from the rest in (6.51) and obtain the results

$$J \neq M: \qquad X(J) = [E - H(J, J)]^{-1} \sum_{K \neq J} H(J, K)X(K) \tag{6.52}$$

$$J = M: \qquad E = \sum_{K} H(M, K)X(K) \tag{6.53}$$

Equations (6.52) are a set of simultaneous equations for the $N - 1$ coefficients $X(J)$ with $J \neq M$, and have a unique solution if E is held fixed. The eigenvalue problem has now become a problem involving the self-consistent choice of E. If the solution of the equation system (6.52) for a given E yields a set of $X(J)$ which reproduce the *same* E when inserted in (6.53), then that E is an eigenvalue and the elements $X(J)$ are the corresponding eigenvector.

6.35 The Brillouin–Wigner method

The initial choice $X(M) = 1$, with all other $X(K)$ zero, will produce a first-order set of $X(J)$. This set, when inserted into (6.53), gives the self-consistent eigenvalue problem traditionally associated with second-order Brillouin–Wigner perturbation theory:

$$E = H(M, M) + \sum_{K \neq M} [E - H(K, K)]^{-1} H(M, K)H(K, M). \tag{6.54}$$

Continuing the perturbative process to get higher-order Brillouin–Wigner eigenvalue problems involves repeated use of (6.52) and (6.53) in a loop structure. However, care must be taken to ensure that all the $X(J)$ calculated in a given application of (6.52) are of the same order, since the *revised* $X(K)$ with $K < J$ will be used in calculating $X(J)$ if (6.52) is simply regarded as an assignment statement. To avoid this difficulty the computed $X(J)$ are first put in a temporary column $T(J)$ and only copied into the $X(J)$ column when all the elements have been computed without any overwriting of the previous generation of $X(J)$ elements.

In principle this Brillouin–Wigner method fits in nicely with the root-finding approach of this book. The basic procedure is to choose an input E, repeat the $X(J)$ calculation a fixed number NT of times for that fixed E and then work out the sum on the right of (6.54). The difference between the input E and the sum (6.54) is the function $F(\mathrm{E})$ which must be rendered zero by the root-finding module. The order of Brillouin–Wigner theory being used is equal to NT + 1, and as NT is increased the zeros of $F(\mathrm{E})$ will tend to the eigenvalues of the matrix H, provided that the off-diagonal elements of H are sufficiently small to permit convergence of the process. A program to carry out this process has been written and used, but it turns out that the program, although of educational value for showing the numerical properties of Brillouin–Wigner perturbation theory, is not as simple or speedy as the other approaches of this chapter.

6.36 The direct iteration procedure

The most direct way to approach the problem is to start with $X(M) = 1$ and simply use equations (6.52) and (6.53) repeatedly as assignment statements, in the hope that the procedure will converge to produce an eigenvalue and its associated eigencolumn. This approach or ones closely similar to it have been known for many years (e.g. Kohn 1949, Boys 1950, Crandall 1951, Nesbet 1965). It essentially involves attempting to solve the equation system (6.52) by a Gauss–Seidel iterative approach. The inverse iteration method described in the results section of FOLDER is based on the same approach. However, it solves the equations (6.52) exactly by using the folding transformation, and thus encounters no convergence problems. For the simple direct iterative method treated here there *can* be a lack a convergence if the off-diagonal elements of the matrix H are not sufficiently small.

The direct iterative method has been used most notably in recent quantum mechanical matrix eigenvalue calculations by Fernandez *et al* (1985) and Fack *et al* (1986); both of these works applied the method to the generalized eigenvalue problem

$$Hx = ESx \tag{6.55}$$

whereas we treat only the ordinary eigenvalue problem (for which S is the unit matrix). The program FOLDER is set up to handle the generalized eigenvalue problem, but it is worth noting that the problem (6.55) can be treated by *any* method designed for the ordinary eigenvalue problem. All that is necessary is to form the matrix $H - \lambda S$ and follow its lowest eigenvalues as λ is varied. $H - \lambda S$ will have a zero eigenvalue whenever λ passes through an eigenvalue of the problem (6.55). The author has found little difficulty in applying this approach, using linear interpolation of the lowest eigenvalue (regarded as a function of λ) in order to get accurate eigenvalues of the generalized problem. The eigencolumn is obviously obtained directly, provided that the technique used is one which produces eigencolumns.

6.37 Relaxation parameters and Rayleigh quotients

When the equations (6.52) are used cyclically the usual procedure is to hold $X(M)$ fixed at the value 1 and input some initial estimate of the Mth eigenvalue. When the off-diagonal elements are small this starting value is usually taken to be $H(M, M)$. If we denote the output from the sum (6.53) by S, then we might simply use the revised choice E = S for the next traversal of the equations. However, as the case of a simple polynomial equation shows (see WYNN) the use of a relaxation parameter is often helpful in producing convergence to some desired root. In the case of the present calculation the relaxation parameter can be introduced at two points. First, the revision of the eigenvalue estimate can be carried out according to the assignment statement

```
E:=RP*S+(1-RP)*E                                    (6.56)
```

with RP being the relaxation parameter. Second, each new element $X(J)$ can be taken as a similar weighted combination of its current value and its modified value as computed using (6.52). Both of these devices have been used in HITTER, and they have been found to help in giving convergence to non-extremal eigenvalues of H. The previous works (cited above) which have used the direct iterative approach in quantum mechanical problems seem to have used the value RP = 1 throughout.

When the matrix H is real symmetric, which ensures that it has only real eigenvalues, several authors have recommended that the revised eigenvalue estimate should be found by using, instead of (6.53), the Rayleigh quotient $R(X)$, defined for a trial vector $X(J)$ by the equation

$$R(X) \sum_J X(J)X(J) = \sum_{J,K} X(J)H(J, K)X(K). \tag{6.57}$$

Kohn (1949) and Nesbet (1965) suggested the direct use of $R(X)$ as the revised E value, since it has the variational property that it gives an eigenvalue estimate in error by order ε^2 when the column $X(J)$ is in error by order ε. A version of HITTER using $R(X)$ in place of the simple sum S has been tested. While it gives more speedy convergence for the eigenvalues, it does not seem to help in giving convergence when the use of S cannot do so. This is presumably because lack of convergence arises from a failure of the Gauss–Seidel process for solving the linear equations; it is the use of this iterative method which both gives the method its simplicity and also limits its range of applicability. A general approach would involve solving the equations more laboriously but exactly by a non-iterative approach, as discussed in connection with the program FOLDER. An obvious feature of the use of the Rayleigh quotient is that it will give a converged eigenvalue estimate *before* the eigencolumn estimates have fully converged. The use of the simple sum S of (6.53) does not give this effect, so that a stopping criterion based on the eigenvalue can be used without impairing the accuracy of the resulting eigencolumn. The variational property of the Rayleigh quotient holds only for the case when the matrix H is symmetric, and its computation for an $N \times N$ matrix takes roughly N times as long as the computation of the simple sum S. Accordingly, the final program HITTER is based on the use of S, although the reader can easily add an $R(X)$ subroutine at the end to check the points which led the author to discard it in favour of the more simple approach based on S.

6.38 The matrix *H*

The computational method of HITTER works satisfactorily for the two extremal eigenvalues of a symmetric matrix, and may also pick out the intermediate eigenvalues if RP is made sufficiently small. However, when the off-diagonal elements are too large the iterations will not converge for the intermediate eigenvalues even when RP is varied. When the matrix H arises as the matrix representation of a quantum mechanical Hamiltonian the relative magnitude of the diagonal and off-diagonal elements depends on the particular set of basis states chosen to set up the matrix. This means that it might be possible to modify the matrix H itself so that the iterative method will work for the intermediate eigenvalues.

The perturbed oscillator Schrödinger equation

$$-\mathrm{D}^2\psi + V_2x^2\psi + V_4x^4\psi = E\psi \tag{6.58}$$

provides a good example of a case in which the matrix H can be adjusted, and is also the one which the perturbation program HYPOSC was designed to treat. The eigenfunction ψ_N of the oscillator Schrödinger equation

$$-\mathrm{D}^2\psi + \beta^2x^2\psi = E\psi \tag{6.59}$$

can be expressed in terms of Hermite polynomials, but the matrix elements of the operator x in the basis ψ_N follow from textbook operator algebra, without detailed knowledge of the ψ_N. The matrix of x is symmetric and has the typical matrix element

$$\langle N|X|N+1\rangle = [(N+1)/2\beta]^{1/2} \tag{6.60}$$

with all other matrix elements zero. In (6.60) the quantum number N is an integer and it labels a state with the eigenvalue $(2N+1)\beta$ for the Schrödinger equation (6.59). The ground state has $N = 0$, whereas the usual convention is to have a lowest index of one for matrices in numerical work. Further, if we want to stick to a basis set of fixed parity we only need to use half of the basis set $\{\psi_N\}$. This leads to an indexation problem in setting up the matrix. If we use the parity indices 0 (for even) and 1 (for odd) then the natural ordering of the matrix elements will pair together the matrix index J and the quantum number N according to the formula

$$N = 2J + P - 2 \tag{6.60}$$

as the reader may confirm. To set up the matrix of the Hamiltonian operator on the left of (6.58) we write the operator in the form

$$(-\mathrm{D}^2 + \beta^2 x^2) + (V_2 - \beta^2)x^2 + V_4 x^4. \tag{6.62}$$

The matrix elements of the first term in parentheses are known, and the matrix elements of the x^2 term follow by taking the square of the X matrix. They are

$$\langle N|X^2|N\rangle = (2N+1)\beta^{-1} \tag{6.63}$$

(which could also be obtained from the virial theorem) and

$$\langle N|X^2|N+2\rangle = [(N+1)(N+2)]^{1/2}\beta^{-1}. \tag{6.64}$$

These two typical elements, together with the statement that the matrix is symmetric, suffice to specify the X^2 matrix. The X^4 matrix then follows by squaring the X^2 matrix, although it is necessary to allow for an 'edge effect'; the last row and column will be in error because there will be a missing contribution from the next state not included in the basis. Using the method outlined above, the matrix of the Hamiltonian operator can be set up for any specified positive β value, and by varying β the relative sizes of the diagonal and off-diagonal matrix elements can be varied. The decomposition (6.62) is the same as that used for HYPOSC, although the parameters used are not defined in the same way. The program BETOSC sets up the matrix and can be used in conjunction with any of the matrix eigenvalue programs of this book. The results section associated with HITTER shows how the choice of β affects the results.

6.39 Algebraic matrix elements

One of the advantages of the direct iterative approach is that (unlike FOLDER) it does not need to store and manipulate *numerical* matrix elements. Provided that there exists an algebraic formula from which the element $H(J, K)$ can be computed whenever it is used, it suffices to retain only the N column elements $X(J)$ when dealing with an $N \times N$ matrix. This advantage was stressed by Fernandez *et al* (1985). It allows large matrices to be treated, even though the computation is slowed down by the repeated recomputation of the same matrix elements. The elements of the matrix of the Hamiltonian operator (6.62) can be given in algebraic form; we have chosen to use a numerical approach, although the algebraic form of the elements of the X^2 matrix is used as part of the numerical work. For two-dimensional problems the algebraic approach becomes almost obligatory. Given a product basis set of type $\psi_N \psi_M$ with only ten ψ_N, the resulting H matrix is of type 100×100. This requires 10^4 elements if a numerical matrix is constructed, but only 10^2 if the $X(J)$ alone are stored numerically. Although the calculation then becomes rather slow, it is remarkable that it can be done on a microcomputer at all. The programs TWODOSC and ZEEMAN of chapter 10 are examples of such calculations.

HITTER. Programming notes

6.40 The relaxation parameter

The assignment statement (6.56) gives the adjusted E value in terms of the current E value and the computed sum S. However, in the programs of this book which use the root-finding approach it is the convention to compute first the shift SH which is to be added to the current E to get the new E. Various tests can then be applied to SH before it is added to E. Inspection of the expression on the right of (6.56) shows that it can be written as E + SH, with

```
SH=(S-E)*RP
```
(6.65)

This shows that RP is simply a multiplying factor which scales the shift up or down from the value which it would take (at RP = 1) for the procedure in which the new E is set directly equal to the computed S. The result (6.65) for SH is the one appropriate for incorporation in a program.

6.41 Displaying the eigencolumn

When the shift value SH is sufficiently small for the E value to have converged, then the $X(J)$ column will be the eigencolumn associated with the eigenvalue E. The procedure adopted in the program FOLDER is appropriate here also; the column can be printed out when the eigenvalue has converged. This is particularly simple for HITTER, whereas FOLDER first has to compute the eigencolumn by a process of inverse iteration.

6.42 The λ parameter

As for the other matrix eigenvalue programs, it is useful to include a multiplying factor λ which is applied to the off-diagonal elements of the matrix H. In the case of HITTER it is anticipated that for sufficiently small λ values all the eigenvalues can be found. It might also be expected that for large λ the two extremal eigenvalues at least should be obtainable, since the Gauss–Seidel iterative process represented by (6.52) is likely to be convergent. A theoretical discussion of this point was given by Nisbet (1972), who treated the case of the generalized eigenvalue problem.

6.43 HITTER. Program analysis and program

Lines 10 to 30 set up the required arrays.

Lines 50 to 80 are the input lines for the matrix H.

Lines 100 to 130 set the off-diagonal multiplier λ, the index M for which the element X(M) is set at unity, the relaxation parameter RP and the initial trial eigenvalue (which is not very critical).

Lines 1000 to 1020 set the initial column of X(J) values.

Lines 1030 to 1110 perform the calculation of the revised X(J) column. Line 1040 ensures that no adjustment is made to X(M), which equals unity throughout. Line 1060 uses a Boolean function factor to omit the K=J term from the sum and also multiplies the off-diagonal elements by the factor λ. Line 1090 guards against a zero denominator in the column calculation. Line 1100 uses the relaxation parameter to control the value of the revised element X(J).

Lines 1200 to 1220 compute the revised eigenvalue estimate E, with the off-diagonal contributions from the elements H(M, K) being multiplied by λ.

Lines 1230 to 1270 compute the shift SH which would arise if the simple substitution E=S were to be made, test for convergence of the iterative E calculation, and apply an actual shift which includes the relaxation parameter RP as a factor. Line 1270 ensures that the column X(J) is *not* reset when the iteration is repeated.

Lines 2000 to 2030 print out the column when the eigenvalue estimate has converged. Control is then passed back to line 100 for the input of a new set of parameters.

The iterative calculation has been put into a module starting at line 1000 so that it can be used (with modifications) as a subroutine in programs to perform other tasks. For example, in a Brillouin–Wigner calculation the column iteration would be carried out a fixed number of times with RP = 1 and the resulting quantity S − E would be returned as a function value to a root-finding module pair such as ROOTSCAN and SECANT. In a Gauss–Seidel solution for a system of linear equations the choices E = 0, $\lambda = 1$ would be made and only the column iteration would be used. S would be replaced by S − Y(J) in line 1100 and line 1040 would be omitted.

```
   7 REM  **********************************
   8 REM  HITTER
   9 REM  **********************************
  10 INPUT "DIMENSION";Q
  20 DIM H(Q,Q)
  30 DIM X(Q): DIM T(Q)
  39 REM  **********************************
  40 PRINT "INPUT MATRIX"
  41 REM  **********************************
  50 FOR J=1 TO Q: FOR K=1 TO Q
  60 PRINT J;K: INPUT X
  70 PRINT X: LET H(J,K)=X
  80 NEXT K: NEXT J
  97 REM  **********************************
  98 REM  INPUT PARAMETERS
  99 REM  **********************************
 100 INPUT "LAMDA";LA
 110 INPUT "INDEX M";M
 120 INPUT "RELAX.PARAM";RP
 130 INPUT "E";E
 996 REM  **********************************
 997 REM  ITERATION
 998 REM  COLUMN CALC.
 999 REM  **********************************
1000 FOR J=1 TO Q
1010 LET X(J)=0
1020 NEXT J: LET X(M)=1
1021 REM  **********************************
1030 FOR J=1 TO Q: LET S=0
1040 IF J=M THEN GO TO 1110
```

```
1050 FOR K=1 TO Q
1060 LET T=LA*H(J,K)*X(K)*(K<>J)
1070 LET S=S+T
1080 NEXT K
1090 LET D=E-H(J,J): IF D=0 THEN LET D=1E-8
1100 LET X(J)=RP*S/D+(1-RP)*X(J)
1110 NEXT J
1197 REM ********************************
1198 REM  ENERGY CALC.
1199 REM ********************************
1200 LET S=H(M,M): FOR K=1 TO Q
1210 LET S=S+M(M,K)*X(K)*LA*(K<>M)
1220 NEXT K
1227 REM ********************************
1228 REM  RELAX.PARAM
1229 REM ********************************
1230 LET SH=(S-E)
1240 IF ABS (SH/E)<1E-8 THEN GO TO 2000
1250 LET E=E+SH*RP
1260 PRINT E
1270 GO TO 1030
1997 REM ********************************
1998 REM  PRINT COLUMN
1999 REM ********************************
2000 PRINT "COLUMN"
2010 FOR J=1 TO Q
2020 PRINT X(J)
2030 NEXT J: GO TO 100
2031 REM ********************************
```

HITTER. Specimen results

6.44 Eigenvalues for BETOSC

To use HITTER together with BETOSC (listed next in this chapter) it is only necessary to add the extra line

```
35 GO SUB 3000: GO TO 100
```
(6.66)

which will use the matrix from BETOSC as the input matrix H. When the β value in BETOSC is chosen reasonably H has small off-diagonal elements and the lowest few eigenvalue calculations converge in around 10 iterations even with RP = 1. Table 6.3 shows the lowest three eigenvalues for the even-parity states of the Hamiltonian $-D^2 + x^2 + x^4$. In this application λ is of course set equal to one. Checking values can be obtained using several of the other programs of this book.

Table 6.3 Eigenvalues with $\beta = 2$.

Q/M	1	2	3
4	1.392 42 30	8.660 879 1	18.112 445
6	1.392 35 19	8.655 203 1	18.062 323
8	1.392 35 16	8.655 050 4	18.057 766
10	1.392 35 16	8.655 050 0	18.057 559

The results of table 6.3 show the correct monotonic descent to the correct eigenvalues which is characteristic of the Rayleigh–Ritz matrix variational method for eigenvalue calculation. At β values of 3 and 4 the same limiting eigenvalues are found and the iterative process still converges quickly. At $\beta = 1$, however, the iterative process converges much more slowly and the limiting eigenvalues are not reached until Q is much larger. Both these effects are due to the fact that the influence of the off-diagonal elements is stronger at $\beta = 1$ than at larger β, and illustrate that the direct iterative approach is best used in conjunction with some procedure which can vary the basis to control the size of the off-diagonal elements.

6.45 BETOSC. Program analysis and program

Lines 3000 to 3020 set up the required arrays. The dimension QM is set equal to Q + 1, to allow for the edge effect in the calculation of the matrix of x^4. Line 3010 is remmed, since the matrix diagonalization program attached to BETOSC will usually have already declared the array H.

Lines 3030 to 3050 are used to set the potential coefficients V_2 and V_4 in the perturbed oscillator Hamiltonian, the quantum number N0 of the state which has the index 1 in the matrix H, and the parameter β which controls the scaling of the basis set of oscillator functions.

Lines 3060 and 3070 precompute two quantities which are used repeatedly in the rest of the computation.

Lines 3100 to 3160 construct the matrix of x^2, using the appropriate analytical formulae. Line 3110 gives the translation formula which puts the computed numbers into the correct positions in the matrix. Line 3150 fills in two equal elements of the symmetric matrix of x^2, which is called A.

Lines 3200 to 3270 form the matrix of $V_2x^2 + V_4x^4$, calling it B, and use the fact that it is symmetric, so that the K loop only needs to go from J to Q. The L sum goes up to QM = Q + 1, to take in the extra contribution which avoids the error due to the edge effect.

Lines 3300 to 3340 put together the final matrix H, using the translation formula in line 3310 and a Boolean function in line 3330 to ensure that only the diagonal elements have the unperturbed contribution added from the term $-D^2 + \beta^2 x^2$ in the Hamiltonian.

Line 3350 is set up to return control to the main program, on the assumption that BETOSC will be a subroutine attached to some matrix eigenvalue program. This use of a subroutine to produce the matrix is a recommended approach, since the choice of problem is then fixed by the subroutine, with the rest of the program left unchanged. The resulting structure is then analogous to that for programs using the root-finding approach, where the particular problem being solved is determined by the function subroutine.

```
2997 REM  *********************************
2998 REM  BETOSC
2999 REM  *********************************
3000 LET QM=Q+1
3010 REM DIM H(Q,Q)
3020 DIM A(QM,QM): DIM B(Q,Q)
3027 REM  *********************************
3028 REM  INPUT PARAMETERS
3029 REM  *********************************
3030 INPUT "V2,V4";V2,V4
3040 INPUT "N0 INDEX";N0
3050 INPUT "BETA";BE
3060 LET GA=S/BE
3070 LET U2=U2-BL*BL
3097 REM  *********************************
3098 REM  FORM X2 MATRIX A
3099 REM  *********************************
3100 FOR J=1 TO QM
3110 LET N=N0+2*J-2
3120 LET A(J,J)=(N+N+1)*GA
3130 IF J=QM THEN GO TO 3160
3140 LET F=GA*SQR ((N+1)*(N+2))
3150 LET A(J,J+1)=F: LET A(J+1,J)=F
3160 NEXT J
3197 REM  *********************************
3198 REM  FORM B=X2*(U2+V4*X2)
3199 REM  *********************************
3200 FOR J=1 TO Q
3210 FOR K=J TO Q: LET S=0
3220 FOR L=1 TO QM
3230 LET S=S+A(J,L)*A(L,K)
3240 NEXT L
3250 LET F=V4*S+U2*A(J,K)
3260 LET B(J,K)=F: LET B(K,J)=F
```

```
3270 NEXT K: NEXT J
3297 REM  *********************************
3298 REM  FORM H=H0+B
3299 REM  *********************************
3300 FOR J=1 TO Q
3310 LET N=N0+2*J-2
3320 FOR K=1 TO Q
3330 LET H(J,K)=B(J,K)+BE*(N+N+1)*(J=K)
3340 NEXT K: NEXT J
3341 REM  *********************************
3350 RETURN
```

6.46 Using FOLDER with BETOSC

To use BETOSC in conjunction with FOLDER we must modify FOLDER slightly by the following addition

```
25 GO SUB 3000: GO TO 55
```

which can be remmed into the program permanently if a variety of matrix-generating subroutines are to be used.

If the Nth energy level of a quantum mechanical system is to be calculated, then the usual Rayleigh–Ritz approach will require the use of a matrix with more than N basis states. The matrix diagonalization which results will give information about all the lower states, but this might not be required if it is only the Nth state which is under investigation. Some recent works (Chang *et al* 1986, Carter and Handy 1987) have explored ways in which the Nth state can be treated by matrix diagonalization techniques which involve much fewer that N basis functions at a time. For the case of the perturbed oscillator problem described by BETOSC it is easy to see how this can be done. If V_4 is very small, then a given unperturbed state N will be linked by the perturbing term V_4x^4 to states between $N \pm 2$ in the first order of perturbation theory, to states between $N \pm 4$ in second order, and so on. The natural way to set up a basis set is to use a basis of unperturbed (or slightly scaled) functions centred on N, with a bandwidth which is gradually increased until the eigenvalues under consideration have converged. The program BETOSC allows the quantum number N0 associated with the matrix index 1 to be varied, while the program FOLDER permits a scan for matrix eigenvalues in any desired region. Table 6.4 shows what happens when the case $V_2 = 1$, $V_4 = 0.0001$ is studied by combining the two programs.

Table 6.4 shows that three energy levels centred on $\mathrm{N0} = 80$ have been found with only nine basis functions, although the omission of the lower basis functions (of which there are roughly 40) means that

Table 6.4 Eigenvalues with $V_2 = 1$, $V_4 = 1E - 4$, $\beta = 1$.

N0, Q	Eigenvalues		
78, 3	157.680 65	161.951 33	166.135 30
76, 5	157.905 33	161.961 07	166.135 30
74, 7	157.914 11	161.961 23	166.009 42
72, 9	157.914 31	161.961 23	166.009 30
70, 11	157.914 31	161.961 23	166.009 30

convergence to the eigenvalues is not necessarily monotonic from above.

For the perturbed oscillator problem described by BETOSC it is clear that the matrix H will be pentadiagonal. For the case of a perturbed Coulomb potential such as $-r^{-1} + \lambda r$ it might appear that no such simplification appears; in a scaled hydrogenic basis all the off-diagonal elements will be non-zero. However, by using an approach based on the theory of the SO(2, 1) Lie algebra, it has been found possible to convert the Schrödinger equation for several perturbed Coulomb potentials into a generalized matrix eigenvalue problem with matrices of banded form (Fernandez *et al* 1985, Fack *et al* 1986).

7 Two perturbation methods

Programs

HYPOSC, RALLY.

7.1 General introduction

Both of the programs of this chapter apply perturbation techniques. The first program, HYPOSC, is one invented and refined by the author in work on the quantum mechanics of the perturbed oscillator, and would serve as a useful example for students who are studying quantum mechanics. The most remarkable feature of the calculation is the way in which the inclusion of one arbitrary parameter K in the algorithm can yield results of such high accuracy for a problem in which the traditional perturbation series is known to have a zero radius of convergence. Two recently discovered features of the approach, the applicability to N dimensions and the production of WKB results, are described, but the interplay of these two features is not yet understood and may provide a useful research project for interested readers. The program RALLY is more traditional, applying perturbation methods to the matrix eigenvalue problem. However, it does not seem that the Padé approximant approach has been applied in this context before, and we found it to give an increase in the range of applicability of the eigenvalue perturbation series. The 'program library' approach advocated in this book makes possible such unorthodox combinations of computational techniques.

HYPOSC. Mathematical theory

7.2 Hypervirial theorems

We start by taking the Schrödinger equation in the form

$$H\psi = -\alpha D^2\psi + V_M x^M \psi = E\psi \tag{7.1}$$

and then work out the mathematical consequences of the requirement that for a stationary state any expectation value of form $\langle\psi|[H, F]|\psi\rangle$ should vanish for an arbitrary operator F. Using the choice $F = x^{N+1}D$ we can break up the evaluation of the commutator into two parts by using the operator identity

$$[H, x^{N+1}D] = [H, x^{N+1}]D + x^{N+1}[H, D] \tag{7.2}$$

which is discussed by Killingbeck (1975) in relation to the Heisenberg equations of motion. The most direct way to evaluate a commutator is to act with it on a dummy operand f; for example

$$\begin{aligned}[H, x^{N+1}]f &= -\alpha[D^2, x^{N+1}]f \\ &= -\alpha D^2(x^{N+1}f) + \alpha x^{N+1}D^2 f \\ &= -\alpha\{N(N+1)x^{N-1}f + 2(N+1)x^N Df\}\end{aligned} \tag{7.3}$$

from which we obtain the operator identity

$$[H, x^{N+1}] = -\alpha N(N+1)x^{N-1} - \alpha(2N+2)x^N D. \tag{7.4}$$

When the term $[H, x^{N+1}]D$ in (7.2) is formed, it will have a component involving the operator $\alpha x^N D^2$. Using (7.1) we replace $-\alpha D^2$ by $E - V_M x^M$, which is correct *provided* that the operand ψ to be used obeys (7.1). Working out the commutator

$$[H, D] = -MV_M x^{M-1} \tag{7.5}$$

and putting together the terms on the right-hand side of (7.2) gives the operator result

$$\begin{aligned}[H, x^N D] = (2N+2)Ex^N - (2N+2+M)V_M x^{N+M} \\ - \alpha N(N+1)x^{N-1}D\end{aligned} \tag{7.6}$$

which is valid for operands ψ which obey (7.1). On taking the expectation values of the operators with respect to a ψ which obeys (7.1) we find for the term involving $x^{N-1}D$,

$$\begin{aligned}\langle x^{N-1}D\rangle &= \int \psi x^{N-1}D\psi\,dx \\ &= \tfrac{1}{2}\int x^{N-1}D(\psi^2)\,dx \\ &= -\tfrac{1}{2}\int \psi^2(N-1)x^{N-2}\,dx\end{aligned} \tag{7.7}$$

after integrating by parts and setting $\psi = 0$ at the boundaries. The final result of the calculation is that the condition that $\langle [H, x^{N+1}D] \rangle$ shall be zero produces the equation

$$\tfrac{1}{2}\alpha N(N^2 - 1)\langle x^{N-2}\rangle = (2N + 2)Ex^N - (2N + 2 + M)V_M\langle x^{N+M}\rangle. \tag{7.8}$$

Clearly, a sum over M is involved if the potential is a series in powers of x instead of the single term $V_M x^M$.

7.3 The perturbation recurrence relation

The most simple case is that in which the potential involves only even powers in x, since then only even N values need be considered in (7.8), as explained by Killingbeck (1987a). However, the hypervirial perturbation method *will* work for mixed-parity potentials such as $x^2 + \lambda x$ (which is also analytically solvable) or $x^2 + \lambda x^3$ (which is not); for such problems all N values $(-1, 0, 1, 2 \ldots)$ have to be considered in the calculation. We take the potential to be $\beta x^2 + \lambda x^4$, which we rewrite as $\mu x^2 + \lambda(x^4 - Kx^2)$, with $\mu = \beta + \lambda K$ *numerically*. We introduce the perturbation series for E and $\langle x^N\rangle$:

$$E = \sum E(M)\lambda^M \tag{7.9}$$

$$\langle x^N\rangle = \sum X(N, M)\lambda^M. \tag{7.10}$$

Substituting these series into the hypervirial relation for the case of a potential $(\mu - K\lambda)x^2 + \lambda x^4$ and picking out the λ^M terms gives the recurrence relation

$$\begin{aligned}(2N + 4)\mu X(N + 2, M)& \\ = (2N + 2)\sum_0^M E(P)X(N, M - P) &+ \tfrac{1}{2}\alpha N(N^2 - 1)X(N - 2, M) \\ + (2N + 4)KX(N + 2, M - 1) &- (2N + 6)X(N + 4, M - 1).\end{aligned} \tag{7.11}$$

These equations are to be used to work out the $X(N + 2, M)$ for $N = 0, 2, 4, \ldots$ and $M = 0, 1, 2, \ldots$, using the starting values

$$X(0, 0) = 1 \qquad X(0, M) = 0 \quad (M > 0) \tag{7.12}$$

with the zeroth-order unperturbed energy

$$E(0) = (2ST + 1)(\alpha\mu)^{1/2} \tag{7.13}$$

where $ST = 0, 1, 2, \ldots$ is the unperturbed oscillator quantum number. To compute a set of coefficients $X(N, M)$ requires knowledge of $E(M)$. When the $X(N, M)$ have been calculated the next energy coefficient $E(M+1)$ follows from the equation

$$(M+1)E(M+1) = X(4, M) - KX(2, M) \tag{7.14}$$

which expresses the fact that $\langle x^4 \rangle - K\langle x^2 \rangle$ must equal $\mathrm{d}E/\mathrm{d}\lambda$.

HYPOSC. Programming notes

7.4 Use of the value $\lambda = 1$

In mathematical work on perturbation theory the Hamiltonian $H_0 + \lambda V$ with $\lambda = 1$ is often studied, with λ regarded as a parameter which turns on the perturbation until it reaches its actual value V. It turns out that the use of $\lambda = 1$ also has a *computational* benefit. The term $E(M)\lambda^M$ in the energy series simply becomes the coefficient $E(M)$; the separate calculation of the factors $E(M)$ and λ^M, followed by their multiplication, can produce spurious overflow or underflow due to either factor, even though their product may be a reasonable number. This effect has been observed in some computations, so the program HYPOSC operates with $\lambda = 1$. To do this it regards the perturbation as $\lambda V_4 x^4$, with $\lambda = 1$; V_4, which is set *numerically* equal to λ, simply becomes an extra multiplying factor in the appropriate term of the recurrence relation.

7.5 Use of the F factor

The $X(N, M)$ and $E(M)$ can range over many orders of magnitude, particularly at $K = 0$, where the perturbation series are strongly divergent. To cut down the range of this variation the program HYPOSC includes a scaling factor F, which can best be understood by rewriting the expansion (7.10) in the form

$$\langle x^N \rangle = \sum X(N, M) F^N \lambda^M \tag{7.15}$$

which essentially involves a redefinition of the coefficients $X(N, M)$. The initial value $X(0, 0) = 1$ is unchanged, but each term in the recurrence relation (7.11) acquires some power of F as an extra factor. The energy equation (7.14) similarly acquires extra factors F^4 and F^2, respectively, in its two terms. The numerical values $F = 2, 4$ or 8 have been found to be satisfactory for a wide range of calculations with the program HYPOSC.

7.6 Calculation of array indices

Inspection of the recurrence relation (7.11) shows that to calculate $X(N+2, M)$ we have to know the elements as far as $X(N+4, M-1)$ in the previous order. Suppose, for example, that we wish to go as far as $E(4)$ with the energy series. Equation (7.14) shows that we need to know $X(4, 3)$, and from (7.11) we see that the following elements would be needed at earlier stages: $X(6, 2)$, $X(8, 1)$, $X(10, 0)$, $X(12, -1)$. The element with $M = -1$ is needed formally (although *numerically* zero) when the equation (7.11) is used to compute the $M = 0$ elements. (The alternative is to have a different calculation at the $M = 0$ level, but HYPOSC sticks to the same loops for every M.) If the example explained above is inspected carefully, it indicates that the calculation of $E(Q)$ requires us to have an array which includes the element $X(N_0, -1)$ where $N_0 = 2Q + 4$. Also, at $M = 0$ the elements up to $X(2Q + 2, 0)$ have to be computed, which means that the index N in (7.11) goes up to $N = 2Q$ in the $M = 0$ computation. However, the maximum N value used can be decreased by two each time the order increases by one, so that only a part of the $X(N, M)$ array is computed; this speeds up the calculation.

Since the index N goes down to zero, while M formally goes down to -1, some shift of the indices must be used for microcomputers which only permit one or zero as the lowest array index. The shift can be permanently written into the indices in the program or can be left as a 'variable constant' SH; in HYPOSC both N and M are shifted by four and the appropriate shifts of four have been incorporated in the array indices throughout the program.

7.7 Calculation of $\langle x^N \rangle$

The values of the $X(N, M)$ for any $N(2, 4, 6, \ldots)$ permit the formation of the sum, equation (7.10) or (7.15), which gives $\langle x^N \rangle$. In HYPOSC the particular quantity $\langle x^2 \rangle$ has been evaluated, but the modification to find some other $\langle x^N \rangle$ is easy to make.

7.8 The calculation of WKB results

In a recent work Killingbeck (1987b) found that the energies and $\langle x^N \rangle$ values which result from the WKB approach to the perturbed oscillator Hamiltonian $-\alpha D^2 + \beta x^2 + \lambda x^4$ will obey the recurrence relation (7.11) provided that the term with the factor α is missed out on the right-hand side. The *unperturbed* energy E_0 is still equal to a multiple of $(\alpha\beta)^{1/2}$

(or $(\alpha\mu)^{1/2}$ if we use the K parameter), since WKB theory gives the exact quantum mechanical energies for a harmonic oscillator potential. The only *numerical* adjustment needed to use HYPOSC to obtain WKB results is to multiply the term involving α by zero when using the recurrence relation; this is done by setting a parameter WF equal to one or zero in the program.

7.9 Variable α and β

In the mathematical literature dealing with the perturbed oscillator it is usual to take the unperturbed Hamiltonian in the form $-D^2 + x^2$. However, the hypervirial theory works equally well for the more general perturbed Hamiltonian $-\alpha D^2 + \beta x^2 + \lambda x^4$ and so HYPOSC allows α, β and λ to be given any required value, with the variation that the input parameters are the coefficients V_2 and V_4 (instead of β and λ).

7.10 The T array and the Wynn algorithm

HYPOSC includes a T ('total') array which stores up the running totals (partial sums) of the energy series at each order. The information can be recalled after each run or, more interestingly, it can be subjected to an analysis by means of Padé approximants, using the program WYNN, which is also included in this collection.

7.11 HYPOSC. Program analysis and program

Lines 10 to 25. The order Q is that of the highest power of λ to be used in the energy series. Line 15 sets the dimensions, as explained in §7.6 and adds four to allow for the shift in indices. Lines 20 and 25 declare E and X arrays and also declare a T array to store the partial sums of the energy series (§7.10). The program will still run if the T array is omitted, but the T array does permit a Padé analysis using WYNN.

Line 30. The scaling factor F is set at two (§7.5), but could be varied. The powers of F are given mnemonic names, F2 for F^2, etc.

Lines 100 to 110. These input lines accept the state quantum number $(0, 1, 2, \ldots)$ and the V and K parameters. The HYPOSC program has the advantage over the programs MOMOSC and SEROSC that the user can specify exactly which state is required (e.g. ST = 5 means the third odd-parity state). The choice K = 0 gives the traditional Rayleigh–Schrödinger perturbation series, while K > 0 gives a renormalized series.

Lines 115, 120. Line 115 sets the value of α in the Hamiltonian $-\alpha D^2 + \beta x^2 + \lambda x^4$, with V2 $\equiv \beta$ and AL $\equiv \alpha$. A separate input line could be used, but we have written AL into the program on the grounds that it is changed less often than the other parameters. The parameter MU $\equiv \mu$ is evaluated in line 115 (see §7.3). The quantities $K\lambda$, $F^2\lambda$, $F^4\lambda$ are all computed here; they appear as coefficients in the recurrence relation (line 225) and are such as to ensure that the factor F^N is included in the definition of the X(N, M) and also that the effective λ for the perturbation calculation is $\lambda = 1$ (§§7.4, 7.5).

Line 125. The parameter WF is set at one for the correct quantum mechanical calculation and at zero for the WKB calculation (§7.8). W is a precomputed coefficient for the α term in the recurrence relation.

Line 130. Here the unperturbed oscillator energy is computed and called E(4), rather than E(0), because of the shift of four in the N and M array indices.

Lines 135, 140. The energy total E is set at E(4) and T(1) is also set at E. The initial condition X(0, 0) = 1 is set (§7.3) but becomes X(4, 4) = 1 because of the index shift. X2 is the series sum representing $\langle x^2 \rangle$ and is initialized to zero. The zeroth-order energy E is displayed.

Lines 200, 205. These set the loop indices. The order index M goes up to Q − 1, since this will mean that λ^Q is the last contribution added to the energy series; E(Q) is found from the X(N, Q − 1) (see §7.3). The N index for a given order M goes up to 2Q − 2M (see §7.6).

Lines 210 to 220. These lines work out the sum over P in the recurrence relation (7.11), as modified to include F factors. Note the index shift of four.

Line 225. This works out most of the sum on the right-hand side of (7.11), with an index shift of four in the arrays and with appropriate factors to allow for the use of the F scaling factor and the use of an effective λ value of unity.

Line 230. The term involving α on the right of (7.11) is added in separately, so that it can be multiplied by zero (i.e. omitted) if the WKB result is required. The WF parameter in line 125 governs this.

Line 235. This evaluates $X(N + 2, M)$ on the left-hand side of the modified form of (7.11), allowing for an index shift of four.

Line 240 ends the N loop with M still fixed.

Line 245. Here the energy equation (7.14) is used, with the extra factors to allow for the scaling factor F, to give the $E(M + 1)$ energy coefficient. Note the index shift and the use of the temporary store S which can be the same as that used in line 215, since the two uses do not interfere.

Line 260 adds on the λ^M contribution to $\langle x^2 \rangle$ (i.e. X2) and adds $E(M + 1)$ to the energy total. Note that λ is effectively one (§7.4) so the value of $E(M+1)$ is *numerically* the same as $E(M+1)\lambda^{M+1}$ for a calculation using a potential term of the type λx^4, with $\lambda = \text{V4}$. The current partial sum of the energy series is stored as an element of the T array. T does *not* need an index shift of four but has its indices chosen so that T(N) is the sum of N terms of the energy series.

Line 265 shows the energy and $\langle x^2 \rangle$ sums so far, and moves on to the next order of perturbation theory (NEXT M). A POKE 23692,3 command added here on the Sinclair Spectrum keeps the screen display scrolling as the output is displayed.

The REM lines included in the program are explained in §7.13.

In HYPOSC the coefficient $X(0,0)$ (named X(4, 4)) is set equal to one in line 135. The M loop starting on line 200 will never work out the X(0, M) with M > 0, although it will *use* their values at some stages. Mathematically these X(0, M) are required to be zero. They acquire this value automatically when the X array is declared on line 25, since a BASIC interpreter always performs this zero initialization, which can be exploited in calculations which use recurrence relations. If the program HYPOSC is set up in another high-level language (e.g. FORTRAN) it may be necessary to include explicit statements which initially set the values of *all* the $X(M, N)$ elements.

```
  7 REM  **********************************
  8 REM  HYPOSC
  9 REM  **********************************
 10 INPUT "ORDER";Q
 15 LET D1=2*Q+8: LET D2=Q+4
 20 DIM E(D2): DIM T(Q+1)
 25 DIM X(D1,D2)
 30 LET F=2: LET F2=F*F
 35 LET F4=F2*F2
 40 LET LL=0
 45 REM INPUT "ANG.MOM";AM
 50 REM LET LL=4*AM*(AM+1)
 51 REM  **********************************
100 INPUT "STATE NUMBER";N0
105 INPUT "V2,V4";V2,V4
110 INPUT "K";K: LET L=V4
115 LET AL=1: LET MU=V2+K*L
120 LET KL=K*L: LET F2L=F2*L: LET F4L=F4*L
125 LET WF=1: LET W=.5*AL*WF
130 LET E=(2*N0+1)*SQR(MU*AL)
132 REM LET E=(4*N0+2*AM+3)*SQR(MU*AL)
135 LET E(4)=E: LET X(4,4)=1
140 LET X2=0: LET T(1)=E
145 PRINT E
```

```
146 REM  ***********************************
200 FOR M=0 TO Q-1
205 FOR N=0 TO 2*(Q-M) STEP 2
210 LET S=0: FOR P=0 TO M
215 LET S=S+E(P+4)*X(N+4,M+4-P)
220 NEXT P
225 LET S=(N+N+2)*S/F2+KL*(N+N+4)*X(N+5,M+3)
    -F2L*(N+N+6)*X(N+8,M+3)
230 LET S=S+W*N*((N*N-1)-LL)*X(N+2,M+4)/F4
235 LET X(N+6,M+4)=S/(MU*(N+N+4))
240 NEXT N
245 LET S=F4L*X(8,M+4)-KL*F2*X(6,M+4)
250 LET E(M+5)=S/(M+1)
255 LET X2=X2+F2*X(6,M+4)
260 LET E=E+E(M+5)
265 LET T(M+2)=E: PRINT E,X2
270 NEXT M
271 REM  ***********************************
```

HYPOSC. Specimen results

7.12 Perturbed oscillator energies

Killingbeck (1985a) gave some numerical results obtained by using HYPOSC for the case $\lambda = 5$ and for state numbers ranging from 0 to 5. Killingbeck (1987b) treated both the cases WF = 1 and WF = 0 which HYPOSC permits, showing how (for the potential x^4) the WKB approximation improves as the state number increases. His results show that the use of 20th order (Q = 20) often suffices to obtain accurate results, provided that K is chosen properly. In general, increasing K pushes the point of 'best convergence' to higher orders, but also improves the number of decimal digits in the 'best-converged' result for E or $\langle x^2 \rangle$.

Table 7.1 is given below with some representative test results, which will serve to check that HYPOSC is working properly. Order 20 is used, since most microcomputers will have sufficient memory space to reach that order; the Sinclair Spectrum 48K computer can go as far as Q = 59 using HYPOSC, although the running time is roughly proportional to Q. At order 20 the WKB results at ST = 0 only converge to five or six digits; however, the poorness of the WKB result for ST = 0 renders further digits of little use.

In the table of test results the WKB HYPOSC energy is shown in parentheses below the quantum mechanical HYPOSC energy. The K value used in each case is shown above the energy values. The roughly linear dependence of the optimum K on ST and V4 enables a good K value to be predicted for intermediate values of ST and V4.

Table 7.1 Test energies obtained using HYPOSC ($V_2 = 1$).

V_4	ST = 0	5	10
	(5)	(10)	(15)
1	1.392 351 6	23.297 441	53.449 102
	(1.250 77)	(23.272 475)	(53.432 882)
	(4)	(6)	(8)
5	2.018 340 7	37.538 815	87.821 861
	(1.7227)	(37.496 344)	(87.794 242)
	(3)	(4)	(5)
10	2.449 174 1	46.729 081	109.772 57
	(2.0613)	(46.675 656)	(109.737 81)

Further results obtained by carrying out a Padé approximant analysis of the series produced by HYPOSC are given in the specimen results section for the program WYNN. Those results show that even when K is somewhat less than the optimum value it is still possible to get a good energy estimate by a Padé analysis of the renormalized series.

7.13 Recent developments

The program HYPOSC as presented and described here is the one used in various calculations reported in the research literature (Killingbeck 1987a, b, 1988b). However, while being prepared for publication, it had some extra REM lines (45, 50 and 132) and two extra variables (AM and LL) added. When the REM lines are activated, they permit calculations for arbitrary angular momentum to be carried out, with the symbol AM being used for l. This was achieved by noting that the inclusion of the extra term $l(l+1)r^{-2}$ in the potential modifies the coefficient of the $\langle x^{N-2}\rangle$ term in the hypervirial relations simply by contributing in just the same way as any other power law potential would. There is also a change in the unperturbed energy; this has to be the energy for an harmonic oscillator state of angular momentum l, and is given in line 132. The program has been checked satisfactorily (with WF = 1) at $l = \pm 1/2$ and 3/2 for the two-dimensional perturbed oscillator. (The general rule for assigning angular momentum values is discussed in the text associated with the program MOMOSC.) The reader is warned, however, that the meaning of the WKB results arising from this recent modification has not yet been fully established.

The theory associated with HYPOSC can clearly be generalized to produce a range of perturbation algorithms which involve slight modifications of HYPOSC without any change in its general structure. For example, a potential of the form

$$V = V_2x^2 + \lambda V_4x^4 + \lambda^2 V_6x^6 \tag{7.16}$$

has been treated by the author (in unpublished work). It turns out that the modified HYPOSC gives accurate energy levels provided that V_6 is suitably small compared with V_4. The use of WYNN helps to improve the results when the variation of K alone is not sufficient to produce high accuracy.

HYPOSC demonstrates a remarkable fact in the development of simple perturbation theory. The study of summability techniques for the perturbed oscillator perturbation series was originally undertaken as a *mathematical* study of divergent perturbation series, with the assumption that as a *numerical* technique perturbation theory could not compete with methods such as that of SEROSC. In fact for a few-term-polynomial perturbation HYPOSC *can* compete numerically and enables expectation values of type $\langle x^N \rangle$ to be calculated along with the energy (in both quantum mechanical and WKB approximations) for any selected state. It illustrates clearly how finding an appropriate algebraic formalism can lead to improved computational algorithms. A similar hypervirial approach is also possible for the perturbed Coulomb potential, although the use of a renormalizing parameter K is not so effective in improving convergence without the extra assistance of WYNN. A hypervirial program for the perturbed potential $-Zr^{-1} + \lambda r$ is given as Appendix 2 of Killingbeck (1985a).

RALLY. Mathematical theory

7.14 Convergence criteria

In the usual textbook presentations of Rayleigh–Schrödinger perturbation theory the perturbed energy coefficients E_n are expressed in terms of sums over the unperturbed eigenstates. This formulation is directly applicable to the matrix eigenvalue problem and we can express the relevant equations in a form adapted to that problem. We suppose that the square matrix A under consideration has the matrix elements $A(J, K)$ and that the off-diagonal elements with $J \neq K$ constitute the perturbation. These off-diagonal elements are to be multiplied by a perturbation parameter λ. At $\lambda = 0$ the eigenvalues $E(J)$ of A are equal to the diagonal elements $A(J, J)$, but will move away from these initial values as λ is increased. It is to be anticipated that for a finite matrix the perturbation series for a given $E(J)$ will converge for λ values less than some critical value (which will usually depend on J). For the simple example of a 2×2 matrix the convergence problem can be investigated analytically. The matrix eigenvalue problem

$$\begin{pmatrix} 1 & \lambda \\ \lambda & 0 \end{pmatrix} \begin{pmatrix} x \\ y \end{pmatrix} = E \begin{pmatrix} x \\ y \end{pmatrix} \tag{7.17}$$

produces a quadratic secular equation with the roots

$$E_{\pm}(\lambda) = \tfrac{1}{2}[1 \pm (1 + 4\lambda^2)^{1/2}]. \tag{7.18}$$

The power series expansion of the square root converges only for values of λ with $|\lambda| < \frac{1}{2}$, even though if λ is real the functions $E_{\pm}(\lambda)$ are well defined for any magnitude of λ. This finite radius of convergence can be traced to the behaviour of the function at complex λ values; at $\lambda = \mathrm{i}/2$ the two branches of the function meet (i.e. the two eigenvalues become degenerate) and each branch has a singularity. For the tridiagonal matrix problem

$$\begin{pmatrix} 1 & \lambda & 0 \\ \lambda & 0 & \lambda \\ 0 & \lambda & 0 \end{pmatrix} \begin{pmatrix} x \\ y \\ z \end{pmatrix} = E \begin{pmatrix} x \\ y \\ z \end{pmatrix} \tag{7.19}$$

the secular equation is a simple cubic equation with the three roots

$$E_0 = 0 \qquad E_{\pm} = \tfrac{1}{2}[1 + (1 + 4\lambda)^{1/2}]. \tag{7.20}$$

The series expansion for the square-root function converges for $|\lambda| < \frac{1}{4}$, but at $\lambda = -\frac{1}{4}$ two eigenvalues meet and there is a singularity in each of the functions $E_{\pm}$. Similar arguments may be used for larger $N \times N$ matrices, and the book by Rellich (1969) includes a detailed discussion of perturbation theory for finite-dimensional spaces.

7.15 The recurrence relations

We study the particular eigenvalue which has the value $A(I, I)$ at $\lambda = 0$, and keep the element $X(I)$ in the associated eigencolumn fixed at the value 1 as λ varies (as for the program HITTER). Remembering that a factor λ is to be incorporated into the off-diagonal elements, we can write the eigenvalue equations in the form

$$A(I, I) + \lambda \sum_{K \neq I} A(I, K)X(K) = E \tag{7.21}$$

and

$$A(J, J)X(J) + \lambda \sum_{K \neq J} A(J, K)X(K) = EX(J) \tag{7.22}$$

(for $J \neq I$). Equation (7.21) incorporates the assumption that $X(I)$ equals one. The perturbation expansions

$$E = \sum E(M)\lambda^M \tag{7.23}$$

and

$$X(J) = \sum C(M, J)\lambda^M \tag{7.24}$$

can be substituted into equations (7.21) and (7.22). Recalling that $E(0)$ is actually $A(I, I)$ we can equate powers of λ on both sides of the equations to obtain the results

$$E(M+1) = \sum_{K \neq I} A(I, K)C(M, K) \tag{7.25}$$

and

$$[A(J, J) - A(I, I)]C(M, J) = \sum_{N=0}^{M-1} C(N, J)E(M, N) - \sum_{K \neq J} A(J, K)C(M-1, K). \tag{7.26}$$

Starting from the initial values $C(0, J) = (J = \mathrm{I})$ the recurrence relation gives the coefficients $C(M, J)$ while (7.25) gives the terms $E(M)$ in the eigenvalue expansion. If the perturbing part of the matrix is taken to include some diagonal contributions, then it is straightforward to add appropriate extra terms with $K = J$ to the sums over K which appear in the equations, leaving the notation $A(J, J)$ to denote the *unperturbed* diagonal element.

RALLY. Programming notes

7.16 The array problem

In several of the programs in this book which are based on few-term recurrence relations it is possible to avoid the use of a large array by storing only the few numbers which are involved in the current application of the recurrence relation; the discarding of previously computed numbers does not impede the progress of the calculation. However, in perturbation calculations of the kind carried out by RALLY (*or* by the hypervirial and inner product methods) there is always a product term such as $EX(J)$ in the algebraic equations. This term inevitably gives a coupling between the eigenvalue and eigencolumn perturbation series. This means that the calculation of Mth-order quantities requires the use of the eigencolumn coefficients from all lower orders, so that these must be stored and cannot be discarded. Fortunately the memory capacity of modern microcomputers is sufficient to store all the computed coefficients and yet allow a high order of perturbation theory to be reached. An alternative way to carry out a perturbation calculation is to use the Brillouin–Wigner approach, in which the E factor in the product $EX(J)$ is *not* expanded as a power series. This cuts down considerably the amount of data storage required, but leads to a more complicated iterative calculation, as explained in the discussion of the program HITTER.

For the case of a symmetric matrix the properties of the Rayleigh quotient mean that if the eigencolumn is known with an error of order λ^{M+1} then the eigenvalue (calculated as the Rayleigh quotient) can be computed with an error of order λ^{2M+2}. To compute $E(11)$, for example, this suggests that knowledge of the eigencolumn coefficients up to the order 5 should suffice. Silverman (1983) described a formalism which allows this result to be exploited directly in the calculation of the $E(M)$ coefficients. A more prosaic approach is simply to work out the Rayleigh quotient explicitly, using the eigencolumn series summed to Mth order. That can be accomplished by adding a short subroutine to the program which calculates the perturbation series.

7.17 The use of $\lambda = 1$

As for the program HYPOSC, the perturbation parameter λ is made effectively equal to one, to avoid using high powers of λ in the perturbation sums. This is accomplished by using directly the products $\lambda A(J, K)$ instead of the $A(J, K)$ as the *off-diagonal* elements in the perturbation equations. This is equivalent to having perturbing matrix elements $\lambda A(J, K)$ with perturbation parameter 1 rather than perturbing matrix elements $A(J, K)$ with perturbation parameter λ.

7.18 RALLY. Program analysis and program

Lines 10 to 50 set up the required arrays. Their dimensions depend on the matrix dimension N and on the maximum order MAX of perturbation theory which it is intended to use. The array T stores the energy sums at each order. It is not essential for the running of RALLY, but is needed if the sequence of partial sums is to be studied using an ancillary program such as WYNN.

Lines 100 to 150 contain the usual matrix input routine.

Lines 200 to 280 allow a specific order of perturbation theory MM less than or equal to MAX to be chosen, together with the perturbation parameter λ which is to multiply the off-diagonal matrix elements. The selected index I sets the particular element A(I, I) which is to give the unperturbed eigenvalue. Lines 230 and 240 set initial values for some scalar quantities, while lines 250 and 280 initialize the coefficient array C(M, J) and the eigencolumn X, using Boolean functions. Note that the indices in the arrays have been increased to avoid zero subscripts in any of the program statements. Thus the initial conditions become E(1) = 0 and C(2, I) =1.

Lines 300 to 420 apply the recurrence relation which works out the C(M, J) coefficients. Two sums are involved in that relation; these are worked out separately (as S and T) and are combined in line 400. Line 320 ensures that the term with K = I is omitted. In line 380 a Boolean factor is used to omit the term with J = K from the sum, and a factor λ is used in the off-diagonal matrix element (as explained in the program notes). Line 410 adds the latest contributions to the X column, with no λ^M factor because of the implicit choice $\lambda = 1$.

Lines 500 to 570 work out the latest eigenvalue term $E(M)\lambda^M$, with $\lambda = 1$, and add it to the eigenvalue estimate E. The Boolean factor in line 520 omits the J = I term from the sum, and also incorporates the factor λ in the off-diagonal matrix elements. The POKE command in line 560 is a Sinclair BASIC command for scrolling the screen display and can be omitted on other microcomputers.

Lines 600 to 650 print out the column as soon as the perturbation calculation has been completed up to the required order.

Lines 700 to 810 work out the Rayleigh quotient RQ as the ratio NUM/DEN. The sums NUM and DEN are worked out within the loops. Note how the J element of the column AX is not stored but is used at once in line 750 to form the product X^TAX. In line 730 the parentheses containing two Boolean functions ensure that the matrix element factors are multiplied by one for the diagonals and by λ for the off-diagonals.

The line pairs 640, 650 and 800, 810 can be used either to stop the calculation or to return to line 200 for a new calculation. The manual commands GO TO 200 and GO TO 700 can be used to control the calculation if the STOP commands are used in the program.

```
 5 REM  **********************************
 6 REM  RALLY
 7 REM  **********************************
10 INPUT "MATRIX DIMENSION";N
20 INPUT "MAXIMUM ORDER";MAX
30 DIM A(N,N): DIM T(MAX+1)
40 DIM C(MAX+2,N): DIM E(MAX)
```

```
 50 DIM X(N)
 99 REM  *********************************
100 PRINT "INPUT MATRIX"
101 REM  *********************************
110 FOR J=1 TO N: FOR K=1 TO N
120 PRINT J;K: INPUT X
130 PRINT " ";X
140 LET A(J,K)=X
150 NEXT K: NEXT J
197 REM  *********************************
198 REM  INPUT PARAMETERS
199 REM  *********************************
200 INPUT "LAMDA";LA
210 INPUT "ORDER";MM
220 INPUT "SELECTED INDEX";I
230 LET E=A(I,I): LET E(1)=0
240 LET T(1)=E
250 FOR K=1 TO N: FOR J=1 TO MM
260 LET C(J,K)=(K=I)*(J=2)
270 NEXT J
280 LET X(K)=(K=I): NEXT K
297 REM  *********************************
298 REM  COEFFICIENTS
299 REM  *********************************
300 FOR M=1 TO MM-1
310 FOR K=1 TO N
320 IF K=I THEN GO TO 420
330 LET S=0: LET T=0
340 FOR J=1 TO M
350 LET S=S+E(J)*C(M-J+2,K)
360 NEXT J
370 FOR J=1 TO N
380 LET T=T+LA*A(K,J)*C(M+1,J)*(J<>K)
390 NEXT J
400 LET C(M+2,K)=(S-T)/(A(K,K)-A(I,I))
410 LET X(K)=X(K)+C(M+2,K)
420 NEXT K
497 REM  *********************************
498 REM  E(M+1) TERM
499 REM  *********************************
500 LET S=0
510 FOR J=1 TO N
520 LET S=S+LA*A(I,J)*C(M+2,J)*(I<>J)
530 NEXT J
540 LET E(M+1)=S: LET E=E+S
550 LET T(M+1)=E
560 POKE 23692,10: PRINT E
570 NEXT M
```

```
597 REM  **********************************
598 REM  EIGENCOLUMN
599 REM  **********************************
600 PRINT "LAMDA=";" ";LA
610 FOR J=1 TO N
620 PRINT J,X(J)
630 NEXT J
640 STOP
650 REM GO TO 200
697 REM  **********************************
698 REM  RAYLEIGH QUOTIENT
699 REM  **********************************
700 LET NUM=0: LET DEN=0
710 FOR J=1 TO N: LET S=0
720 FOR K=1 TO N
730 LET S=S+A(J,K)*X(K)*(LA*(K<>J)+(K=J))
740 NEXT K
750 LET NUM=NUM+X(J)*S
760 LET DEN=DEN+X(J)*X(J)
780 NEXT J
790 LET RQ=NUM/DEN: PRINT RQ
800 STOP
810 REM GO TO 200
811 REM  **********************************
```

RALLY. Specimen results

7.19 Rayleigh–Schrödinger mode

By modifying a few lines of the program it is possible to include the calculation of the Rayleigh quotient within the M loop which starts at line 300, and to print the values of E and RQ side by side on the screen at each order. This was done for a 4×4 test matrix with all its off-diagonal elements equal to one and its diagonal elements given by $A(J, J) = J$. For λ values up to 0.3 the perturbation calculation clearly converges to eight digits when taken up to order 40, but the RQ sequence converges more quickly, needing only about half as many terms as the perturbation sum. To get the accurate eigencolumn, of course, it is necessary to wait until the perturbation series has converged; the point is that the RQ value gives a good estimate of the eigenvalue even using a low-order eigencolumn estimate which has not fully converged. At and beyond $\lambda = 0.4$ the series does not give converged results, nor does the sequence of RQ values. However, good eigenvalues *are* obtained by performing a Padé analysis of the sequence of partial sums T(M) of the eigenvalue series. This extends up to roughly $\lambda = 0.8$ the region of usefulness of the eigenvalue series.

7.20 Brillouin–Wigner mode

In an *ad hoc* experiment, it was found possible to run RALLY in a manner which applies Brillouin–Wigner perturbation theory. The temporary modifications used were as follows.

1. New lines:

```
225 INPUT"TRIAL E";ET
335 GO TO 370
795 NEXT M
796 LET ET=(E+ET)/2: GO TO 230
```
(7.27)

2. REMS introduced at the start of lines 560, 570, 620, 650. PRINT RQ in line 790 changed to PRINT E, RQ, and A(I, I) replaced by ET in line 400.

With the above changes the use of a high-order (such as 50) leads to converged eigenvalues at λ values up to about $\lambda = 0.7$, roughly about as far as the Padé analysis of the Rayleigh–Schrödinger eigenvalue series was able to go. For larger λ the Brillouin–Wigner approach can still give the extremal eigenvalues but fails for the intermediate ones. In all cases the RQ values give accurate eigenvalues considerably before the eigenvalue process has converged.

7.21 Rayleigh iteration mode

If E is changed to RQ in line 796 of the temporary modifications, then the calculation will proceed by forcing ET and RQ to be equal (at least if convergence is obtained). This process converges more quickly than the Brillouin–Wigner calculation. A study of the relevant algebra reveals that the calculation is very similar to the process of Rayleigh iteration, with the novelty that a perturbation formalism is being used to compute the effect of the matrix $(A - ET)^{-1}$ on a unit column $e(I)$. The quantity RQ is then the Rayleigh quotient for the resulting column $(A - ET)^{-1}e(I)$, and yields a better ET value for the next iteration; however, RQ is evaluated for the next cycle long before the column calculation has converged! In a full Rayleigh iteration process the eigencolumn is obtained also, and the operand is not held at $e(I)$ but changes from one cycle to the next. The perturbation version given here gives at $\lambda = 0.5$ the accurate eigenvalues 0.735 985 1, 1.806 177 5, 2.896 338 5 and 4.561 498 9. These have the correct sum 10, as required by the diagonal sum of the matrix, and agree with the results of FOLDER for this problem.

8 Finite difference eigenvalue calculations

Programs

RADIAL, FIDIF.

8.1 General introduction

Both of the methods of this chapter are based on finite difference simulations of the Schrödinger differential equation, and so are amenable to the Richardson extrapolation methods described in chapter 2. However, the two programs are intended to illustrate different approaches. RADIAL is a method with h^2 error, but illustrates how the calculation of expectation values can be incorporated into the energy level calculations. It also uses a factorization of the Schrödinger radial equation which removes the centrifugal potential energy term from the potential. The program FIDIF, on the other hand, uses a more traditional approach for the one-dimensional Schrödinger equation, but then applies ideas of perturbation theory to improve the accuracy of the computed energy values. The third method included in the flexible FIDIF program has been found to be more accurate than the widely used Numerov method, particularly for the excited states of smooth potentials. Both of the methods of this chapter are based very strongly on an appropriate algebraic analysis, as the text makes clear.

RADIAL. Mathematical theory

8.2 The finite difference approximation

The method described here is not the most powerful of the finite difference methods available, but is quite adequate to give accurate energies for the low-lying levels of the Schrödinger equation. We derive the relevant equations in a somewhat unorthodox manner (Killingbeck 1977) which makes them applicable to the radial equation for states of any angular momentum as well as to the usual one-dimensional Schrödinger equation for problems such as the perturbed harmonic oscillator. One unusual feature of the resulting equation is the absence of the centrifugal $l(l+1)r^{-2}$ term which appears in the traditional radial equation as derived in quantum mechanics textbooks.

The Schrödinger equation is taken in the form

$$-\alpha\nabla^2\psi + V\psi = E\psi \tag{8.1}$$

where V is the potential function, which is assumed to depend only on the radial coordinate r. To pick out states of angular momentum l it is usual to write ψ as a product of a radial factor and a spherical harmonic function of the angular variables. We use instead the postulated form

$$\psi = \mathcal{Y}_l W(r) \tag{8.2}$$

where $\mathcal{Y}_l$ is a solid harmonic (i.e. a product of the factor r^l and an angular function) and W depends only on r. $\mathcal{Y}_l$ has the useful mathematical properties

$$\nabla^2\mathcal{Y}_l = 0 \qquad r\frac{\partial\mathcal{Y}_l}{\partial r} = l\mathcal{Y}_l \tag{8.3}$$

which simplifies the algebra when the product (8.2) is substituted into the Schrödinger equation (8.1). The result of that substitution is the following equation (after removing a factor $\mathcal{Y}_l$)

$$\alpha^{-1}r(V - E)W = r\mathrm{D}^2W + (2l + 2)\mathrm{D}W \tag{8.4}$$

where D denotes differentiation with respect to r. This equation does not contain a term of type $l(l+1)r^{-2}$, but the price paid for this simplification is the appearance of a first derivative term which involves l.

To convert the differential equation into a finite difference form we use the simple central difference formulae of lowest order

$$W(r + h) - W(r - h) = 2h\mathrm{D}W(r) \tag{8.5}$$

$$W(r + h) + W(r - h) - 2W(r) = h^2\mathrm{D}^2W(r). \tag{8.6}$$

These approximations convert (8.4) to the form

$$rF(r)W(r) = [r + H]W(r + h) + [r - H]W(r - h) \tag{8.7}$$

where

$$F(r) = 2 + \alpha^{-1}h^2[V(r) - E] \tag{8.8}$$

and $H = (l + 1)h$. If a constant step length h is used then we can set $r_N = Nh$ and use the integer N to label the W values at the discrete points r_N. With $M = l + 1$, (8.8) then takes the form

$$NF(r_N)W(N) = [N + M]W(N + 1) + [N - M]W(N - 1) \quad (8.9)$$

with $F(r_N)$ being computed by setting $r = r_N$ in (8.8).

8.3 The shooting approach

The equation (8.9) is a three-term recurrence relation. It is clear by inspecting (8.8) and (8.9) that they could be combined and rearranged to produce a matrix eigenvalue equation of form $Mx = Ex$, where M is a tridiagonal matrix and x is the eigencolumn with the $W(N)$ as its elements. This way of treating the finite difference versions of the Schrödinger equation has been adopted by several authors, but we shall adopt instead a shooting method. This method is more intuitive in that it involves proceeding along the r axis and tracing out the wavefunction W step by step. The idea is to vary the trial energy E until the W value at the outer boundary is zero (for the most common case of homogenous Dirichlet boundary conditions). This means that the shooting method falls into the family of root-finding methods which were discussed generally in chapter 2. Although the equations derived here apparently refer to states of angular momentum l for spherically symmetric potentials in three dimensions, they are actually of wider application. For example, with the choices $l = -1$ and $l = 0$ the equations hold for states of even and odd parity in one dimension (provided, of course, that the potential has even parity). For the even-parity ($l = -1$) case the factor N cancels throughout equation (8.9) and is omitted. It can also be shown (Hikami and Brezin 1979, Killingbeck 1985d) that states of angular momentum Λ for isotropic potentials in D dimensions can be handled by setting $l = (D + 2\Lambda - 3)/2$ in the equations for the three-dimensional case, so that our formalism could be applied to such states, after some modification, since we take l to be an integer in the program as presented here.

8.4 Expectation values

In order to apply the ideas of chapter 2 to calculate expectation values from the finite difference recurrence relation, we must be able to calculate the rate of change of $W(r = L)$ with respect to various parameters, where $W(r = L) = 0$ is the test which is applied to locate the energy eigenvalues. Thus, if the recurrence relation (8.9) is differentiated with respect to E, with due allowance for the explicit dependence of $F(r_N)$ on E, we obtain the result

$$NF(r_N)W_E(N) - \alpha^{-1}h^2NW(N) = [N+M]W_E(N+1) + [N-M]W_E(N-1) \tag{8.10}$$

where $W_E(N)$ denotes the derivative of $W(N)$ with respect to E.

To proceed further in the calculation of the expectation value of some function $U(r)$ we add a dummy term $\mu U(r)$ to the potential $V(r)$ and then differentiate the recurrence relation (8.9) with respect to μ. The result is

$$NF(r_N)W_\mu(N) + \alpha^{-1}h^2U(r_N)NW(N) = [N+M]W_\mu(N+1) + [N-M]W_\mu(N-1) \tag{8.11}$$

where $W_\mu(N)$ is the derivative of $W(N)$ with respect to μ. The three recurrence relations (for W, W_E and W_μ) have most of their coefficients in common, but both the W_E and W_μ sequences are coupled to the W sequence by means of a term which explicitly contains $W(N)$ as a factor.

To apply the recurrence relations derived above there are two possible approaches. If only the energy eigenvalues are required it is sufficient to use (8.9) to compute successive W values up to the value on the outer boundary ($r = L$) at which we wish to impose the boundary condition $W(L) = 0$. If a trial energy E is the input to the subroutine, then the output will be the $W(L)$ value for that E; this means that the $W(L)$ program can be used in conjunction with a root-finding program to locate the energy levels. An alternative approach is to use the W_E recurrence relation in parallel with the W recurrence relation, so that the ratio $W(L)/W_E(L)$ can be used directly in Newton's method to find the roots of $W(L)$. The recurrence relation for W_μ, which yields $W_\mu(L)$, can be used to compute $\langle U \rangle$ from the ratio $W_\mu(L)/W_E(L)$, as explained in the general discussion of chapter 2. However, using the W_μ recurrence relation all the time will slow down the calculation; it is quite sufficient to use it only after an eigenvalue has been located, to find $\langle U \rangle$ for the corresponding eigenfunction.

The finite difference approach has the advantage that the potential $V(r)$ can be any smooth function, whereas methods such as MOMOSC and SEROSC are better suited to simple polynomial potentials. However, there is a discretization error associated with finite difference methods. As explained in chapter 3, it is necessary to use two or more values of the step length h, together with Richardson extrapolation, in order to obtain energies and expectation values of high accuracy. The simple method described here has a discretization error series with a leading term of h^2 type.

RADIAL. Programming notes

8.5 Initial conditions

To start off a recurrence relation which involves three terms at a time it is usually necessary to assign values to the first two terms and then compute all the later terms. Because of the linearity of the Schrödinger equation and of the recurrence relations arising from it, it suffices to set $W(0)=0$ at the origin and to set $W(1)$ equal to any fixed number, which can be chosen to be h if we wish W to have unit slope at the origin. For the special case of the recurrence relation (8.9) a simple alternative would be possible. If we suppose that $W(l)$ is finite it follows that it does not contribute in the calculation of $W(l+2)$ since it has a zero multiplying factor in the equation (8.9) when we set $N = l+1$. This means that we could start at $N = l+1$, giving $W(l+1)$ some fixed finite value, with all subsequent $W(N)$ being scaled up or down by choosing $W(l+1)$ appropriately. In fact we chose to keep the W calculation standard by starting always at $r=0$. It is amusing to note that the choice $W(0)=0$ is actually wrong for s states, yet does not change the computed energies.

8.6 Storage requirements

Since the quantity required for locating the energy eigenvalue is the last $W(N)$ value (i.e. the one on the outer boundary) it follows that the other $W(N)$ values do not need to be stored as an array, unless we wish to retain information about the wavefunction over the region considered. It suffices to store only three W values at a time, since only three values are used in the W recurrence relation. In a similar manner only three W_E and three W_μ need to be stored. In actual computations the number of steps used is typically 100 or 200, and so the necessary storage space would be for something less than a thousand real numbers even if all the quantities were stored. That is well within the capacity of modern microcomputers; nevertheless, it is still useful to see how the problem might be treated in a manner which minimizes the storage requirements.

8.7 Node counting

The roots E_n of the equation $W(L) = 0$ will give approximate energy levels for the Schrödinger equation (approximate because of the h^2-type error), but it is useful to know *which* energy level (first, second, etc) we have found when the calculation produces an E value. For the radial or the one-dimensional Schrödinger equation the number of zeros in the wavefunction increases by one as we pass from one energy level to the next higher one. The relevant algebra to establish this result can be found in various textbooks (e.g. Kemble 1958, Killingbeck 1975). If the grid of r values used in the finite difference approach is sufficiently fine to have several strips within each local de Broglie wavelength of the wavefunction, then a count of the number of sign changes in the sequence $W(N)$ can be used to give the number of zeros in the eigenfunction, which then indicates the number of the state being considered. For example, the ground state of the radial equation will have $W = 0$ at $r = 0$ and $r = L$, where we use the test $W(L) = 0$ to locate the energy levels. The nth excited state will have n extra nodes between 0 and L, and so a count of these will indicate to which state a calculated energy value belongs.

8.8 The choice of L

The standard choice of outer boundary condition for those cases of the Schrödinger equation which are analytically solvable is $W(\infty) = 0$. Although in the Hill–series approach (SEROSC) it is possible to simulate this condition in a computation, we clearly cannot compute out to infinity in a finite difference method which uses a finite step length. Fortunately, the bound states of most potentials of interest can be found by using the condition $W(L) = 0$ for some finite L and then increasing L until the computed energy levels stabilize at what we can take to be their asymptotic values corresponding to the limit $L \to \infty$. In terms of the wavefunction this limit will be reached when L is such that the true wavefunction $\psi(r)$ is negligibly small at and beyond $r = L$. In terms of the energy, if the potential function $V(r)$ at and beyond $r = L$ is much greater than the energy of the states being considered, then use of the condition $\psi(L) = 0$ should suffice to simulate the condition $\psi(\infty) = 0$, since only an exponential tail of the exact wavefunction will survive in the region $r > L$. The direct way to check that the asymptotic value of the energy has been reached is to increase L and repeat the calculation. Another check is to note that according to the virial theorem the quantity

$$\langle r\,\mathrm{d}V/\mathrm{d}r + 2V\rangle - 2E \qquad (8.12)$$

should be zero when the boundary is effectively at infinity. This quantity has the dimensions of energy, and for simple one-dimensional problems it can be shown to equal $L\partial E/\partial L$.

8.9 The index *I*

The calculation of the function $W(L)$ is performed in a function-evaluating subroutine. In the initial scan only W is needed; in the detailed root finding W_E is also needed; in the expectation value calculation W_μ must be added. Thus the subroutine may be required to evaluate one, two or three functions as the integration proceeds along the r axis. We use an index I, which each calling part of the program feeds to the subroutine and which specifies the number of functions which are to be evaluated in the subroutine.

8.10 RADIAL. Program analysis and program

Line 5 sets up small E and R arrays, just in case we wish to store the results for different h values for later use with an automatic Richardson extrapolation routine. RADIAL as displayed here does not mention these arrays again; they could be used (after the STOP at line 430) by means of a manual instruction such as LET E(1)=E:LET R(1)=EU which stores the current values. These stored values will not be lost if a manual GO TO 50 is used to vary h.

Lines 10 and 20 give the potential $V(r)$ and the function $U(r)$ for which the expectation value is required, and so will be varied from problem to problem.

Line 30 gives the coefficient α which multiplies $-\nabla^2$ in the kinetic energy operator. This also varies from problem to problem.

Lines 40 to 100 are for various inputs and precompute several quantities which are held fixed throughout the program. The traditional letter L for angular momentum is used, with the *distance* L being called RM.

Line 100 sets the starting energy and the energy interval DE to be used in the initial scan to locate a root.

Line 110 sets the indicator I to 1 for use in the subroutine at 1000.

Lines 200 to 230 scan along the E axis until $W(L)$, produced by the subroutine at 1000, changes sign between E and E + DE.

Lines 300 to 370 use the values of F and G produced by the subroutine with the index I set at two to activate both the F and G calculations in the subroutine. The shift according to Newton's formula is evaluated in line 320, checked for excessive magnitude in line 330 and added to E in line 340. Line 310 is used to show E, F and the node count NZ (i.e. number of zeros). Line 350 checks for convergence of the computed E value and jumps to the expectation value calculation when E has been found.

Lines 400 to 430 set I = 3, to activate all three available calculations in the function subroutine, and print out the expectation value of the function specified in line 20.

Lines 1000 to 1030 set the initial values of various quantities. W is the wavefunction, WE is W_E, and WU is W_μ in a mnemonic type of notation.

Lines 1040 to 1100 compute the $W(N)$ along the r axis. Line 1090 counts the number of sign changes in W along the axis. Note that in line 1100 we directly pass to W1 and W2 without introducing a quantity W3.

Lines 1110, 1150 use the index I, sent by the calling routine, to switch on or off the calculations of WE and WU which proceed in parallel with that of W.

Lines 1120 to 1200 These two sections give WE and WU when required. Note that in lines 1140 and 1190 it is W1 (not W2) which is used. The *current* W value is required by the mathematical theory; that *should* be the W2 value, but in line 1100 the value of W2 has been moved on to the W1 location. In an earlier version of the program W2 was wrongly used, with the result that the $\langle U \rangle$ values showed an h (rather than an h^2) error law. This error was detected by performing several test computations with varying h.

Lines 1210 to 1240 prepare the function values for return to the calling routine. W2 is the W value at the end of the integration region (i.e. at $r = L$). When I = 1, the value of G is not returned, but will have the formal value zero, as set in line 1010.

```
 1 REM  **********************************
 2 REM  RADIAL
 3 REM  H=-AL*W"+V(R)
 4 REM  **********************************
 5 DIM E(5): DIM R(5)
 9 REM  **********************************
10 DEF FN V(R)=R*R
20 DEF FN U(R)=R*R
21 REM  **********************************
30 LET AL=1
40 INPUT "RMAX";RM
50 INPUT "STRIP NUMBER";N0
60 LET H=RM/N0: LET A=H*H/AL
70 INPUT "ANG.MOM";L
80 LET M=L+1
```

```
100 INPUT "E,DE";E,DE
110 LET I=1: GO SUB 1000
197 REM  ********************************
198 REM  ROOTSCAN
199 REM  ********************************
200 PRINT E,NZ,F
210 LET ES=E: LET FS=F
220 LET E=E+DE: GO SUB 1000
230 IF F/FS>0 THEN GO TO 200
297 REM  ********************************
298 REM  ROOT
299 REM  ********************************
300 LET I=2: GO SUB 1000
310 PRINT E,NZ,F
320 LET SH=-F/G
330 IF ABS SH>DE/2 THEN LET SH=DE/2*SGN SH
340 LET E=E+SH
350 IF ABS (E/ES-1)<2E-8 THEN GO TO 400
360 LET ES=E: GO SUB 1000
370 GO TO 310
399 REM  ********************************
400 LET I=3
410 GO SUB 1000
420 PRINT EU
430 STOP
997 REM  ********************************
998 REM  WFN. CALCULATION
999 REM  ********************************
1000 LET W1=0: LET W2=1
1010 LET WE1=0: LET WE2=0
1020 LET WU1=0: LET WU2=0
1030 LET R=0: LET NZ=0
1031 REM  ********************************
1040    FOR N=1 TO N0-1
1050 LET R=R+H
1060 LET V=FN V(R)
1070 LET FV=2+A*(V-E)
1080 LET S=N*FV*W2+(M-N)*W1
1090 LET NZ=NZ+(S/W2<0)
1100 LET W1=W2: LET W2=S/(M+N)
1110 IF I=1 THEN GO TO 1200
1117 REM  ********************************
1118 REM  ⟨U⟩ CALCULATION
1119 REM  ********************************
1120 LET S=N*FV*WE2+(M-N)*WE1
1130 LET WE1=WE2
1140 LET WE2=(S-A*N*W1)/(N+M)
1150 IF I=2 THEN GO TO 1200
```

```
1160 LET U=FN U(R)
1170 LET S=N*FV*WU2+(M-N)*WU1
1180 LET WU1=WU2
1190 LET WU2=(S+A*N*W1*U)/(N+M)
1200   NEXT N
1209 REM ***********************************
1210 LET F=W2: LET G=WE2
1220 IF I<>3 THEN GO TO 1240
1230 LET EU=-WU2/G
1240 RETURN
1241 REM ***********************************
```

RADIAL. **Specimen results**

8.11 The Coulomb potential

We show some results for a standard potential, the Coulomb one (with $\alpha = 1/2$). Table 8.1 shows some results for the case L = 10 and the ground state of the potential $V(r) = -r^{-1}$. The exact results in the limit $L \to \infty$ are $E = -1/2$, $\langle r \rangle = 3/2$. The table shows how Richardson extrapolation is needed to get accurate results; we should note here that increasing L to 12 produces the result E = −0.499 999 8 and $\langle r \rangle =$ 1.499 997, which are even closer to the correct asymptotic values (we round the final results to seven digits).

Table 8.2 shows results of lower accuracy, obtained by using only two N0 values (50 and 100), to illustrate how the long range of the Coulomb potential causes the wavefunctions of the 2s and 2p states to extend over a large distance. The exact results for $L \to \infty$ are $E = -1/8$ (for both states), with $\langle r \rangle$ equal to 6 for the 2s state and 5 for the 2p state.

When a Coulomb potential perturbed by a term of form λr or λr^2 is

Table 8.1 $V = -r^{-1}$, $\alpha = 1/2$, L = 10. Richardson extrapolated values are given in parentheses

N0	E	$\langle r \rangle$
50	−0.495 096 74	1.510 011 1
(50, 100)	(−0.499 975 02)	(1.499 910 8)
100	−0.498 755 45	1.502 435 9
(100, 150)	(0.499 996 51)	(1.499 932 6)
150	−0.499 444 93	1.501 045 2
(50, 100, 150)	(−0.499 999 2)	(1.499 935)

Table 8.2 Results for 2s and 2p states, with $V = -r^{-1}$, $\alpha = 1/2$, using N0 = 50 and N0 = 100 with Richardson extrapolation

L	State	E	$\langle r \rangle$
15	2s	−0.12450	5.867
20	2s	−0.12498	5.991
15	2p	−0.12477	4.927
20	2p	−0.12499	4.996

treated, as in the spherical Zeeman effect problem (Killingbeck 1981c) or in charmonium calculations (Killingbeck and Galicia 1980), then the use of large L values is not necessary, since the potential cuts off the wavefunction strongly as r increases. For the radial harmonic oscillator problem, with $V = r^2$ and $\alpha = 1$, RADIAL gives accurate levels for the first few states of each angular momentum using only 50 and 100 steps. The virial theorem result $E = 2\langle r^2 \rangle$ is obeyed very closely as soon as L is such that the boundary potential L^2 is large compared with the energy being calculated.

8.12 A comment on singular potentials

RADIAL uses one of the most simple possible finite difference methods. Although it gives a discretization error of order h^2, it does have the advantage that it works well even for potentials which have a singularity at $r = 0$. Alternative methods such as the Numerov one, although they give an h^4-type error for non-singular potentials, revert to an h^2 error law when the potential has a singularity. It is also possible to vary the step length throughout the integration in some finite difference methods. RADIAL uses a fixed h because for that case the extra calculations used to obtain the derivatives and the expectation values take a form which is particularly simple, being as close as possible in form to those used for the wavefunction calculation.

For the case of the singular potential $Cr^{-6} + r^2$ Jamieson (1983) gave a careful and illuminating analysis of the way in which the discretization error of the finite difference eigenvalues depends on h. He showed the presence of various exponential terms in the error, with the error tending to h^2 type as h becomes sufficiently small. For such singular potentials the use of a matrix diagonalization approach is complicated by the presence of divergent matrix elements.

FIDIF. Mathematical theory

8.13 The three-term recurrence relation

For a small step length h the central difference quantity

$$\delta^2 \psi(x) = \psi(x + h) + \psi(x - h) - 2\psi(x) \tag{8.13}$$

has the following expansion in powers of h:

$$\delta^2 \psi(x) = h^2 \mathrm{D}^2 \psi(x) + (h^4/12)\mathrm{D}^4 \psi(x) + \dots . \tag{8.14}$$

This result is used to provide a finite difference approximation to the Schrödinger equation

$$-\alpha \mathrm{D}^2 \psi + V\psi = E\psi \tag{8.15}$$

in one dimension. If the h^4 and higher-order terms in (8.14) are neglected, we may combine (8.14) and (8.15) to obtain the result

$$\delta^2 \psi(x) = \alpha^{-1} h^2 [V(x) - E]\psi(x) \tag{8.16}$$

which may then be expanded using (8.13) to produce a three-term recurrence relation

$$\psi(x + h) = [2 + \alpha^{-1} h^2 (V - E)]\psi(x) - \psi(x - h). \tag{8.17}$$

This equation can be used in a shooting calculation in the same way as in the program RADIAL. For one-dimensional symmetric potentials such as the perturbed oscillator potential $x^2 + \lambda x^4$, the eigenfunctions can obey two different boundary conditions at $x = 0$. Odd-parity states have $\psi(0) = 0$, $\mathrm{D}\psi(0) \neq 0$, while even-parity states have $\psi(0) \neq 0$, $\mathrm{D}\psi(0) = 0$. The two starting ψ values must thus be varied to pick out states of the desired parity, and the integration process need only be carried out from $x = 0$ to $x = L$, where we impose the outer boundary condition $\psi(L) = 0$. When the resulting finite difference method is used to calculate eigenvalues, the energies have a discretization error of order h^2, just as for RADIAL.

8.14 Obtaining h^4 error

By working through equations (8.13) to (8.17) again, we see that the kinetic energy operator has the expansion

$$-\alpha \mathrm{D}^2 = -\alpha h^{-2} \delta^2 - (\alpha/12) h^2 \mathrm{D}^4 + \dots . \tag{8.18}$$

It follows that neglecting terms beyond the first one on the right is equivalent to introducing a perturbing term of order h^2 into the Schrödinger equation. From first-order quantum mechanical perturbation theory we thus expect an error of leading order h^2 in the computed energy, which is what is observed. However, as pointed out by Killingbeck (1979a), that first-order perturbation theory also allows us to calculate and eliminate the h^2 error term in the energy; this produces a method which gives a leading energy error of h^4 type while still only using a three-term recurrence relation. The relevant algebra involves working out the first-order energy shift caused by a perturbing term $h^2\mathrm{D}^4$ and showing that it equals the expectation value of the local operator $\alpha^{-2}h^2(V - E)^2$. The final result is that equation (8.17) can be modified to obtain energies with an error of leading order h^4. The only change required is to define the quantity

$$f(x) = \alpha^{-1}[V(x) - E] \tag{8.19}$$

(which can be computed as a named quantity in a computation) and then to use in the square bracket of the recurrence relation (8.17) the quantity $2 + fh^2 + (1/12)h^4f^2$ instead of $2 + fh^2$. This idea can be taken further (Killingbeck 1986a): for the case of a constant potential V the rest of the series in powers of h can be given analytically so that the three-term recurrence relation is exact. Unfortunately, for variable V it turns out that a term involving $\mathrm{d}V/\mathrm{d}x$ appears in the h^6 term, which takes the form $h^6[f^3/360 + (\mathrm{D}V)^2/240]$.

If the root-finding approach to the problem is used, then the shooting calculation becomes a function subroutine which returns the value of $\psi(L)$ when the calling routine sends a value of E. It is possible to construct a version of FIDIF which computes expectation values, as does RADIAL. The relevant theory is given by Killingbeck (1985c); however, the version of FIDIF given here is kept as simple as possible in order to permit the testing of various possible functions of f in the square bracket of the recurrence relation (8.17). In particular, the subroutine does not work out the gradient G (as RADIAL does) and so must be used in conjunction with an external root-finding routine.

8.15 Initial and boundary conditions

To start off the recurrence relation at $x = 0$ for odd-parity states we can set $x = h$, with $\psi(0) = 0$ and $\psi(h)$ some finite fixed number. For even-parity states we know that $\psi(-h) = \psi(h)$; this means that we can set $\psi(0)$ equal to some non-zero value, compute $\psi(h)$ as though $\psi(-h)$ were zero and then divide it by two. Thus the case of even-parity states is the only one requiring any special treatment. If the potential $V(x)$ is not of even parity, then the integration cannot start at $x = 0$; it must proceed from some large negative x to some large positive x, with the starting condition exactly that which would be used for an odd-parity state. Indeed, it is a good test to apply this 'full-axis' calculation to the case of an even-parity potential to check that it still picks out correctly the energies found by a 'half-axis' calculation starting at $x = 0$. When an even-parity potential is of double-well type, with even- and odd-parity levels separated by a small splitting, it is of particular advantage to have distinct 'half-axis' calculations for the two types of state.

8.16 Forwards and backwards shooting

One of the defects of the forwards shooting procedure (from $x = 0$ to $x = L$, for example) is that the wavefunction values at large x are not obtained accurately. This arises because the method is based on the properties of the three-term recurrence relation, which mimic those of the Schrödinger differential equation quite well. The solution to the recurrence relation contains two components; at large x values one of these (the dominant one) grows exponentially, while the other decreases exponentially. If the trial E were exactly equal to the desired eigenvalue, E_n, then the dominant solution would have a zero coefficient. However, in general there will be a dominant contribution, with an amplitude proportional to $(E - E_n)$ if this energy difference is small. This means that only a slight variation in E gives a large change in $\psi(L)$; the result is that the roots E_n are found accurately but the wavefunction near $x = L$ is not known accurately.

There are various ways to deal with this problem. One widely used method is to shoot from both ends and match the solution at some intermediate x (Cooley 1961, Talman 1980). An approach suggested by Killingbeck (1987c) involved a forwards shooting process to find $\psi(L - h)$, with $\psi(L)$ held at zero, followed by a backwards shooting process to obtain $\psi(x)$ throughout the entire region. We should note here that the method for calculating expectation values, which is discussed in chapter 2 and used in the program RADIAL, was introduced partly to avoid this problem of an imprecise wavefunction, since it avoids calculating the integrals which appear in the traditional definition of an expectation value. To make it possible to investigate this wavefunction problem experimentally, it is useful to make a finite difference program capable of either forwards or backwards shooting, with the imposition of either Dirichlet ($\psi = 0$) or Neumann ($D\psi = 0$) boundary conditions at either end. This degree of flexibility has been incorporated in FIDIF along with the capability of using different coefficients in the three-term recurrence relation.

Despite the comments about the wavefunction problem (or, perhaps, because of them) FIDIF is presented here in a form which retains only three ψ values at a time, using the name W for ψ. To investigate the wavefunction it is only necessary to add a PRINT instruction to show the current W value on the screen or printer, or to store the W value as the W(N) element of an array which can be declared at the start of the program. The point about the program as presented here is that it is intended to be applicable even on a microcomputer with limited array storage capacity.

8.17 The Numerov method

One of the most commonly used and cited methods for handling the Schrödinger equation is the Numerov method, which for smooth potentials has a leading error term of h^4 type in its energy estimates. It reverts to an h^2 error for singular (e.g. Coulomb) potentials, although this seems to be little noted in the literature. We can obtain the relevant equation by noting that $D^4\psi$, which appears in equation (8.18), is the second derivative of $D^2\psi$, which obeys the Schrödinger equation (8.15). Equation (8.18) can thus be written in the alternative form

$$\delta^2[\psi - (h^2/12)f\psi] = h^2 f\psi \tag{8.20}$$

with f defined in (8.19). This result is a compact form of the Numerov equation, which is applied by expanding the δ^2 term according to the definition (8.13). If, however, we define a new function

$$\chi = [1 - (h^2/12)f]\psi \tag{8.21}$$

then (8.20) takes the simple form

$$\delta^2\chi = h^2 f[1 - (h^2/12)f]^{-1}\chi \tag{8.22}$$

$$= [h^2 f + (h^4/12)f^2 + (h^6/144)f^3 + \ldots]\chi. \tag{8.23}$$

The first two terms on the right are exactly the same as those obtained by our earlier perturbation theoretical argument, although χ differs from ψ by a leading error of order h^2. The point is that as abstract computational formulae, with the boundary conditions that ψ (or χ) equals zero, the Numerov method and the simple method of §8.14 give the *same energies* (to order h^4). The perturbation theoretical approach (Killingbeck 1979a, 1986a) also shows that an exact finite difference formula would contain an $h^6 f^3$ coefficient of 1/360, whereas the Numerov method gives a coefficient 1/144. Method 2 as used in FIDIF uses a zero coefficient, which usually gives slightly more accurate energies.

8.18 The three methods of FIDIF

The program FIDIF can be used in three modes, each one of which uses a particular factor *FV* in the three-term recurrence relation

$$\psi(x + h) = FV(x)\psi(x) - \psi(x - h) \tag{8.24}$$

although the structure of the program is sufficiently simple for the user to insert a 'pie-filling' *FV* of any flavour! All three methods incorporated in FIDIF as presented here use the function f as defined in equation (8.19). *Method 1* has

$$FV = 2 + h^2 f \tag{8.25}$$

and gives a leading energy error of order h^2, which means that it is often necessary to use three different h values plus Richardson extrapolation to obtain seven or eight decimal digits of accuracy.

Method 2 uses the improved formula which was explained in terms of perturbation theory in the preceding discussion:

$$FV = 2 + h^2 f + (h^4/12)f^2. \tag{8.26}$$

The argument given above shows that the resulting method is very similar to the Numerov method as far as energy level calculations are concerned, giving a leading error of h^4 type. However, the *wavefunction* as obtained by using (8.26) in (8.24) still has an error of h^2 type; the perturbation method is simply an elegant way of extracting from it a better energy estimate. To obtain a wavefunction with h^4 leading error it is necessary to exploit the link with the Numerov method and use the relation (8.21), that is *divide* the calculated wavefunction by $1 - (h^2/12)f$ at every x to produce a new wavefunction corrected for the h^2 error.

Method 3 relies on a study of the simple case of a constant potential $V(x)$. Depending on the sign of f, which is the same as that of $V - E$, the function *FV* can be seen to represent the first three terms of the series expansion of either $2\cos\theta$ or $2\cosh\theta$, with $\theta^2 = h|f|$. For a constant potential the use of one of these functions (which one depending on the sign of f) gives a 'sum to infinity' of the series in $h^2 f$ which makes the recurrence relation fit exactly to the Schrödinger differential equation. For a varying potential it seems reasonable to use the *local V* value in the constant potential formulae. However, as V varies so will the sign of $V - E$ and f, which means that the appropriate program steps will have to test the sign of f and then work out a cos or cosh function. This is easy to do, although it might involve constructing the cosh function in terms of the exp function on some microcomputers. In FIDIF, however, we have used a simple but good approximation to the required function; this involves no special functions and takes the same form for both possible signs of f. The basic idea comes from some algorithms originally designed to produce cos and cosh functions using a simple electronic calculator (Killingbeck 1981a); a similiar approach has recently been applied in the calculation of certain elliptic functions (Coqueraux *et al* 1990).

The fundamental mathematical result required is an angle-doubling formula

$$2\cos 2\theta = (2\cos\theta)^2 - 2 \tag{8.27}$$

which *also* holds for the cosh function. The choice $\theta^2 = h^2 f$ produces the FV of (8.26) as the low-order power series approximation to the desired function (cos or cosh, depending on the sign of f). Applying the series (8.26) to $\theta/2$ first and then using (8.27) gives a more accurate value for the function than is obtained by using θ directly in (8.26). The change $\theta \to \theta/2$ is produced by the change $f \to f/4$, so the final form of the required algorithm is (in terms of assignment statements)

$$g := h^2 f/4 \tag{8.28}$$

$$Y := 2 + g + g^2/2 \tag{8.29}$$

$$FV := Y^2 - 2 \tag{8.30}$$

where g and Y are arbitrary names for the intermediate quantities. This short algorithm gives a good simulation of the local constant potential approximation for the three-term shooting method used in FIDIF, particularly since the effective angle θ (being proportional to h) is usually small.

8.19 The propagator approach

Constant potential approximations have also been used in the propagator method of shooting (Devries and George 1980). In that method the pair $(\psi, \mathrm{D}\psi)$ at $x + h$ is derived from the pair $(\psi, \mathrm{D}\psi)$ at x by the action of a 2×2 matrix, the elements of which depend on f and h. The propagator method is based on the representation of the second-order Schrödinger equation in terms of two coupled first-order equations; if we denote the wavefunction by W and its gradient by G we see that the Schrödinger equation (8.15) can be transformed into the following pair of equations (with D denoting $\mathrm{d}/\mathrm{d}x$):

$$\mathrm{D}W = G \tag{8.31}$$

$$\alpha \mathrm{D}G = (V - E)W. \tag{8.32}$$

Various finite difference approximations can then be used to propagate the pair (W, G) along the x axis, with the special advantages that the step length h can easily be varied as the calculation proceeds, and that boundary conditions involving both W and G can be handled directly. Even so, the propagator formulae have to be quite complicated before they can give energy values as accurate as those given by the very simple formulae used in FIDIF. The simple propagator method SPLEEN of chapter 4 gives only h^2 accuracy.

8.20 Excited state calculations

Much of the literature concerned with finite difference eigenvalue calculations has concentrated on the h dependence of the error, with the intuitive notion that an h^4 or h^6 method is 'better' than an h^2 one. Less attention has been paid to how the error depends on the energy, yet this is an important consideration if calculations are to be carried out for 10 or so excited states as well as for the ground state. The methods used in FIDIF all involve various powers of the quantity $h^2(V - E)$ and it is possible to do a theoretical analysis to establish how the energy eigenvalue error depends on E and h, at least in the asymptotic limit $E \to \infty$. Method 1 turns out to have a leading error term of type E^2h^2, while method 2 (like the closely related Numerov method) has a leading error term of type E^3h^4. Method 3 is more difficult to analyse; it has a leading error term of form $F(E)h^4$, where $F(E)$ is a function of E which, according to our current estimates, grows not much faster than linearly with E. For rising non-singular potentials, such as perturbed oscillator potentials, method 3 represents a remarkable advance over the Numerov method and yet requires a very simple algorithm. The methods of FIDIF *can* be applied to singular potentials, such as those involving $-r^{-1}$ or $l(l+1)r^{-2}$ but will then lose their higher-order accuracy in h and revert to an h^2 type of leading error term, with an E dependence which we have not yet investigated.

FIDIF. Programming notes

8.21 The range of x

Although most calculations will be carried out for $0 \leqslant x \leqslant L$, it may be necessary to use other ranges of x to try the effect of shooting backwards instead of forwards. This means that the program must be set up to use arbitrary end points X1 and X2; the choice X1 > X2 must then produce a *negative* step length H, so that the instruction X = X + H will lead to a backwards traversal of the region from X1 down to X2. It is also necessary to compute the wavefunction one step *beyond* each boundary, since the condition that W shall show reflection symmetry across a boundary may sometimes need to be applied.

8.22 An economical starting procedure

We shall take the case X1 = 0, X2 = L to explain the idea of this procedure. For an odd-parity state the choice W1 = 0, W2 = 1 for the first two wavefunction values is clearly appropriate, if W1 is the $\psi(0)$ value. For an even-parity state, however, the property $\psi(-h) = \psi(h)$ must be imposed. We can *still* retain the choice W1 = 0, W2 = 1, *if* we associate W1 with $\psi(-h)$ instead of $\psi(0)$ *and* remember to divide the computed W3 value (i.e. $\psi(h)$) by two to allow for the reflection symmetry about the origin. The input parameter P1 (parity at X1) is chosen to be zero for even parity and one for odd parity. In the basic recurrence relation (8.24) the current X value, for which V is evaluated, is that for which the wavefunction is ψ(X), which is named W2 in the program. To ensure that the value W2 = 1 refers to X = 0 for P1 = 0 (even parity) and to X = H for P1 = 1 (odd parity) we use the first X value H*P1, or X1 + H*P1 for the most general case. To ensure that the first computed W3 value, that is ψ(H), is divided by two for the even-parity case we can use a dividing factor such as 1 + (X = 0) which incorporates a Boolean function.

The program FIDIF uses very slight variations of the above ideas; the point is that only simple functions of the parity parameter P1 are needed to pick out the even- and odd-parity starting conditions, with W1 and W2 held fixed. This simple procedure, based on a study of the relevant mathematical equations, only evolved gradually. The first version of FIDIF used a separate starting routine, involving some ten program lines, to set up the initial values of the variables for the even-parity case. Although the version presented here involves a mathematically redundant division by one at each X value beyond X = 0, it is more compact and simple in form than the original version.

8.23 The boundary conditions

At the end point X = X2 the condition W(X2) = 0 is the one most commonly applied. However, to impose the symmetry condition [W(X2 − H) − W(X2 + H)] = 0 we have to calculate for one step beyond the boundary, which actually involves working out the potential *at* the boundary. If we use the brief notation W and G for the two quantities mentioned above, then the quantity which we require to be zero can be written as W*P2 + G*(1 − P2), where P2 (0 or 1) is the parity parameter at the X2 boundary.

One point involving some mathematical subtlety should be made here. Use of the test G = 0 is appropriate to pick out even-parity states (for even-parity potentials) when backwards shooting from X1 = L to X2 = 0 is used. If, however, the test G = 0 is applied when X2 is an *outer* boundary, it is not exactly equivalent to the use of the homogeneous Neumann boundary condition $D\psi = 0$. We can see this by noting that for functions which obey the Schrödinger equation (8.15) we have the following result: if $D\psi(X) = 0$ then

$$\psi(x + h) - \psi(x - h) = (h^3/3\alpha)\psi(x)(DV(x)) + O(h^5). \quad (8.33)$$

This means that the condition G = 0 will differ from the condition $D\psi = 0$ by a leading term of order h^3. However, if we use $G/(2h)$, that is a proper central difference estimate of the gradient, the leading error in the boundary condition involves $h^2 DV(X)\psi(X)$, and should in principle simply add a small h^2 error term into the error series for the energy. If the outer boundary X2 is sufficiently remote for ψ to be very small there, this h^2 contribution will be of negligible magnitude.

The reasoning above also indicates that in the case where X2 is out in the asymptotic region for the wavefunction we would expect to get the same energy levels for homogeneous Dirichlet and Neumann boundary conditions. For the case X2 = 0, with an even-parity potential, the factor $DV(x)$ is zero in (8.33), and there is no mismatch between the Neumann boundary condition and the one which requires reflection symmetry.

8.24 Wavefunction scaling

As presented here, FIDIF uses the initial values 0 and 1 for the wavefunctions W1 and W2. If the region of integration is chosen to be too long, then X2 will be very far into the asymptotic region of the wavefunction. The dominant solution of the recurrence relation (§8.24) might then grow so large that it gives computer overflow before X2 is reached. There are several ways to deal with this, which we can list as follows.

1. Decrease the length of the integration region; this will not change the calculated energy if X2 *is* well into the asymptotic region.
2. Use a much smaller initial W2 value (e.g. IE − 20). This sometimes suffices, but the growth of the dominant solution can be sufficiently rapid to defeat this approach.

3. Use a redefined wavefunction WS (W scaled) such that

$$\mathrm{WS}(Nh) = \mathrm{W}(Nh)K^{N-P_1} \tag{8.34}$$

WS2 still has the initial value 1, but the recurrence relation is slightly modified to take the form

$$K^2\mathrm{WS3} = K(FV)\mathrm{WS2} - \mathrm{WS1}. \tag{8.35}$$

This approach is very effective, with K typically between 1 and 100, although care must be taken to modify appropriately the formulae which incorporate the boundary conditions.

4. As soon as the W value exceeds some fixed large value (e.g. IE20) at X = XC, say, proceed with the rest of the calculations *as though* X2 had been reached. This procedure assumes that the asymptotic region has already been reached (as evidenced by the large value of the dominant solution), so that proceeding further to X2 would not change the calculated energy. The difficulty is that XC must be stored and also used as the upper limit for subsequent values of the trial energy. As the root finder closes in on the root the overflow problem usually disappears, since the amplitude of the dominant solution will be diminished by a factor of 1E8 or more.

8.25 FIDIF. Program analysis and program

Line 10 gives the potential as a user-defined function.

Line 20 sets the coefficient α in the kinetic energy operator $-\alpha\mathrm{D}^2$.

Lines 30 to 50 set the limits of integration, the end parities and the number of strips. A manual GO TO 50 can be used to change the number of strips.

Lines 60 and 70 work out the step length H (which will be negative if X2 < X1) and precompute the fixed number H^2/α which is used many times in the calculation.

Lines 80 and 90 allow the user to pick one of the three methods explained in the text, with the parameter K being set at zero for method 1 and at 1/12 for methods 2 and 3. Note the use of the Boolean function.

Line 100 sets the initial trial energy E0 and the increment DE by which E is to be increased in the ROOTSCAN module.

Line 110 asks for the scaling factor SF and works out its square SF2. SF is usually set at one. If the program stops because of overflow in the shooting subroutine, then the manual instruction GO TO 100 is used, with SF set at gradually increasing values (2, 4, 6, etc) until the calculation proceeds without overflow. If SF is made too large, the resulting function values may come out as zero, which will probably halt the program because of arithmetic problems in the root-finding module.

Lines 200 to 270 constitute the ROOTSCAN module, which is standard except that it prints NZ (the number of nodes in the wavefunction) as well as E and F.

Lines 300 to 360 constitute the root-finder module SECANT. Note that line 346 can have the number 2E − 8 modified to allow for the number of digits of accuracy given by the microcomputer. Note also that the line assumes that in an IF statement of the form IF A THEN B:C both B and C will be carried out if condition A is satisfied. Most BASICS obey this rule, but a modification will be necessary for those which do not.

Lines 1000 to 1150 carry out the finite difference shooting calculation.

Lines 1010, 1020, 1090, 1140 incorporate the parameters P1, P2 and SF in accord with the ideas explained in the programming notes.

Line 1050 is an IF statement of the same type as that in line 346, and so the remarks made about that line are also applicable here.

Lines 1060 to 1080 apply method 3 (explained in §8.18) if I = 3.

Line 1120 applies the idea explained in §8.22, dividing by two only for N = 0, that is the case in which reflection symmetry is imposed on the wavefunction at the starting coordinate X1.

Line 1140 picks out either the wavefunction $\psi(X2)$ or the difference $\psi(X2 + H) - \psi(X2 - H)$, according to the value of the parity parameter P2, as explained in §8.23.

Line 1150 returns the appropriate function value F to the ROOTSCAN or SECANT modules, which send a trial E value to the shooting subroutine.

```
6 REM  ***********************************
7 REM  FIDIF
8 REM  H=-AL*W"×V
9 REM  ***********************************
10 DEF FN V(X)=X*X
11 REM  ***********************************
20 LET AL=1
30 PRINT "X LIMITS,X1,X2"
35 INPUT X1,X2
40 PRINT "PARITIES P1,P2"
45 INPUT "0 EVEN,1 ODD";P1,P2
50 INPUT "STRIP NUMBER";N0
60 LET H=(X2-X1)/N0
70 LET A=H*H/AL
80 INPUT "METHOD 1,2 OR 3";I
```

```
  90 LET K=(I<>1)/12
 100 INPUT "E0,DE";E0,DE
 110 INPUT "SF";SF: LET SF2=SF*SF
 197 REM  *********************************
 198 REM  ROOTSCAN
 199 REM  *********************************
 200 LET E=E0
 210 GO SUB 1000: PRINT E;"Z";NZ,F
 220 LET F1=F: LET E1=E
 230 LET E=E+DE
 240 GO SUB 1000: PRINT E;"Z";NZ,F
 250 LET F2=F: LET E2=E
 260 LET R=F2/F1
 270 IF R>0 THEN GO TO 220
 297 REM  *********************************
 298 REM  SECANT
 299 REM  *********************************
 300 LET E=E2-DE+DE/(1-R)
 305 LET FS=F2: LET ES=E2
 310 GO SUB 1000: PRINT E,F
 330 LET GI=(E-ES)/(F-FS)
 340 LET ES=E: LET FS=F
 345 LET SH=-F*GI
 346 IF ABS (SH/E)<2E-8 THEN LET E=E2: GO TO 210
 350 LET E=E+SH
 360 GO TO 310
 997 REM  *********************************
 998 REM  WAVEFN. CALC.
 999 REM  *********************************
1000 LET W1=0: LET W2=1
1010 LET X=X1+H*(P1-1): LET NZ=0
1019 REM  *********************************
1020    FOR N=P1 TO N0
1030 LET X=X+H: LET V=FN V(X)
1040 LET AV=A*(V-E)
1050 IF I<>3 THEN LET FV=(1+K*AV)*AV+2: GO TO 1090
1051 REM  *********************************
1060 LET Y=AV/4
1070 LET Y=(1+K*Y)*Y+2
1080 LET FV=Y*Y-2
1081 REM  *********************************
1090 LET W3=(SF*W2*FV-W1)/SF2
1100 LET NZ=NZ+(W3/W2<0)
1110 LET W=W1: LET W1=W2
1120 LET W2=W3/(1+(N=0))
1130    NEXT N
1140 LET F=P2*W1+(1-P2)*(W2*SF2-W)
1150 RETURN
```

FIDIF. **Specimen results**

8.26 Comparison of methods 2 and 3

For the case of potentials of form $\mu x^2 + \lambda x^4$ it is easy to pick out any particular excited state and obtain an accurate energy for it by using the perturbation program HYPOSC. This gives us a supply of test energies with which to see how the errors of methods 2 and 3 compare as the energy increases; this point was discussed in §8.20. Table 8.3 shows some results for the Hamiltonian $-D^2 + x^4$. Only five or six digits are shown; later digits are in error for both methods at the fixed N0 value of 50 which is used. Both methods give an h^4-type leading error term. The use of N0 = 50 and N0 = 100, with Richardson extrapolation, gives results for method 3 which are accurate to eight digits for the states shown. For method 2 it is necessary (at least for the higher states) also to use N0 = 150 and take into account the small h^6 error term to obtain eight-digit accuracy.

The results of the table show how well method 3 works for this potential, as it does in general for smooth monotonic potentials. Method 2, which gives results very close to those of the Numerov method, has an error which grows much more rapidly with E.

Table 8.3 Even-parity energies for $-D^2 + x^4$, with X1 = 0, X2 = 5. The number of steps used is N0 = 50

Method 2	Method 3	Accurate
1.0604	1.0604	1.0604
7.4557	7.4557	7.4557
16.2623	16.2617	16.2618
26.5310	26.5284	26.5285
37.9308	37.9231	37.9230
50.2752	50.2567	50.2563
63.4423	63.4044	63.4030
77.3463	77.2761	77.2732
91.9240	91.8035	91.7981

8.27 Extrapolation for method 1

Method 1, for which results are not shown in table 8.3, is obviously much less accurate than methods 2 and 3 at N0 = 50. However, it *does* give accurate results if used with Richardson extrapolation, which takes a particularly simple form for an error series with terms which are all multiples of h^2. We can explain this by using the results of method 1 for one of the even-parity energy levels of $-D^2 + x^4$ which appears in table 8.3. Table 8.4 displays the results for four N0 values, together with the extrapolated results.

The extrapolation rule can be stated in words as follows: proceed as though the error is of h^2 type, but use the two outer N0 values which were needed to produce the pair being extrapolated. For example, the number 77.273 196 in the table is obtained by h^2 extrapolation using the N0 values 100 and 200, since these led to the two preceding numbers in the table. Chapter 3 also explains the rule.

Table 8.4 Results for method 1, with X1 = 0, X2 = 5 and the same potential as used for table 8.3.

N0				
		(77.2)	(77.273)	
50	74.392 232			
		809 09		
100	76.558 740		151	
		740 13		199
150	76.956 114		196	
		734 00		
200	77.094 927			

8.28 Expectation values

Although FIDIF is not set up to give expectation values directly it will give reasonably accurate values just by adding small terms of form $\lambda U(x)$ to the potential and observing the energy shift to estimate $\langle U \rangle$. As an unusual example table 8.5 shows estimates of $\psi^2(x)$ obtained for $V = x^2$ by using the perturbing function

$$U(x) = (\lambda/2\mathrm{H})*(\mathrm{X} > \mathrm{X0} - \mathrm{H}/2)*(\mathrm{X} < \mathrm{X0} + \mathrm{H}/2) \qquad (8.36)$$

which simulates a delta function at X = X0. The factor 1/2 is necessary for even-parity potentials, since a pair of delta functions at $\pm x$ would give an expectation value $2\psi^2(x)$.

The results of the table show that the method works adequately even without the use of several N0 values, although they can be used to improve the estimated values. The N0 value of 128 was chosen both to give accurate energies with method 3 and to ensure that the X0 value is on a grid point. The analytic values in the table are, of course, rounded to show the first digit of disagreement.

Table 8.5 Results obtained using the $U(x)$ of equation (8.36), $V = x^2$, X1 = 0, X2 = 8, N0 = 128.

(X0, λ)	Energy	Approx. ψ^2	Analytic ψ^2
1, 0.01	1.002 073 3		
		0.207 553	0.207 554
1, −0.01	0.997 922 2		
2, 0.1	1.001 019 6		
		1.010 335	1.010 333
2, −0.1	0.998 952 6		
3, 1	1.000 064 0		
		0.000 070 1	0.000 069 6
3, −1	0.999 923 7		

8.29 Final comments

The literature on finite difference methods is vast; the author's collection of *selected* reprints on that topic contains several hundred items. Accordingly, the few works mentioned here are ones which directly relate to the ideas used in the developments of FIDIF.

The perturbation theory approach (§8.14) has been applied to improve the accuracy of simple band theory calculations in one dimension (Killingbeck 1980). To avoid the need for scaling the wavefunction (§8.24) it is possible (with a little complication about boundary conditions) to set up the finite difference equations in terms of the ratios $\psi(x+h)/\psi(x)$. This was done by Killingbeck (1977) in an early version of RADIAL and also by Johnson (1977) in a computationally elegant treatment of the Numerov method.

The speciment results for FIDIF showed how methods 2 and 3 differ in the energy dependence of their errors, although both have an h^4 error in terms of h. Andrew (1986) gave a mathematical analysis of the h^4E^3 error law for the Numerov method, pointing out that the method of Andrew and Paine (1985) allows this to be reduced to an $h^4E^{3/2}$ error. The idea behind the correction method used can be appreciated intuitively by recalling that the quantity $V - E$ plays a central role in the theory. For high-energy states we can hope to get an estimate of the error by setting $V = 0$ in the theory. However, that gives us the 'particle in a box' system of quantum mechanics, for which many of the equations arising in the finite difference theory can be solved *analytically*. When this analytic error estimate for $V = 0$ is included in the calculation for the full V it leaves a residual error which increases less rapidly with E than the error for the uncorrected method. The principle behind this approach is presumably capable of further extension, for example by using an oscillator V as the reference potential, or by obtaining the error for the reference potential *computationally*. We estimate that the error for the Numerov method with the Andrew and Paine correction is roughly the same as that for the method 3 of FIDIF.

The correction method outlined above refers to the case of a non-singular potential. For a singular potential all the usual h^4 methods revert to being of h^2 error type, just like the method of RADIAL. However, Buendia and Guardiola (1985) proposed a method which retains the h^4 error law for the Numerov method for cases in which (as in Coulomb potential problems) the potential is singular at the origin. Their approach may be adaptable for the methods of FIDIF, although the author has not yet investigated that point.

The author tends to favour the shooting approach to finite difference eigenvalue problems. However, some workers prefer to treat the problem in terms of matrix eigenvalue techniques. As noted in the discussion of SEROSC, the recurrence relation approach can often be seen to be an economical way of locating the eigenvalues of some equivalent matrix problem. Porter and Reiss (1986) gave a careful and informative account of the relationship between the shooting and the matrix approaches to finite difference eigenvalue problems. Recently, Frantz *et al* (1989) have pointed out the value of 'wrong-way' recursion for the accurate determination of eigenvalues (see §8.16).

9 Recurrence relation methods

Programs

SERAT, SEROSC, MOMOSC.

9.1 General introduction

The three programs of this chapter use recurrence relation methods to calculate the eigenvalues of the Schrödinger equation for the case in which the potential is a sum of a few power law terms. The programs SERAT and SEROSC are based on a power series expansion of the wavefunction. However, they take advantage of the author's research on the links between the power series and Hill determinant methods, and are dual-purpose Hill – series programs. The mathematical trick which makes this possible, forwards nested multiplication, is explained in §9.5. The 'obvious' choice of the parameters β and γ in the algorithm is explained to be wrong in §9.7; this is a remarkable example of a counter-intuitive result revealed by computational experiment, and the 'obvious' choice appears to have misled several previous workers in this area. The program MOMOSC is based on the use of moment recurrence relations. The technique used will actually permit perturbation calculations for problems (such as the two-dimensional oscillator) where the hypervirial method fails, although that application is not given in this book. The function $F(E)$ produced in MOMOSC is of tan-like type, with singularities between its zeros; this makes the root-finding process a little more difficult.

SERAT and SEROSC. Mathematical theory

9.2 The perturbed Coulomb potential

SERAT is designed to calculate energy levels of the Schrödinger radial equation

$$-D^2\psi + [l(l+1)r^{-2} - Zr^{-1} + V_1r + V_2r^2]\psi = E\psi \quad (9.1)$$

for arbitrary angular momentum l. The angular momentum need not be an integer; Killingbeck (1988c) used a sequence of l values varying from 0 to 1 to show how the 1s state energy passes over smoothly to the 2p state energy as l is increased for a weakly perturbed Coulomb potential. Varying l slightly permits the calculation of the expectation value $\langle r^{-2}\rangle$ from the energy shift formula

$$\Delta E = (2l+1)\langle r^{-2}\rangle \Delta l \quad (9.2)$$

in accord with the method used at several points in this book.

The use of a power series approach in solving the Schrödinger equation is possible for any potential which has a rapidly converging power series, for example the screened Coulomb cosine potential $V = \exp(-\mu r)\cos\lambda r$ (Killingbeck 1988c). However, SERAT is kept as simple as possible and deals with the perturbed Coulomb potential of equation (9.1), although the reader will find it easy to add r^3 and r^4 terms once the structure of the program has been understood. Power series methods for energy level calculation have been used by many authors (e.g. Secrest *et al* 1962, Killingbeck 1981b, Barakat and Rosner 1981). In recent developments links between the power series approach and the Hill determinant approach have been extensively investigated both theoretically and computationally (Chaudhuri 1985, Killingbeck 1986b, Banerjee 1978, Hautot and Nicolas 1983, Znojil 1976).

The potential of equation (9.1) is sufficiently general to encompass several applications of recent interest. The simple models of the charmonium system in elementary particle theory use potentials of that type (Killingbeck and Galicia 1980), with the r^{-1} term being called the gluon term and the other terms representing a confining potential. For the case in which the $-Zr^{-1}$ term is taken to be the nuclear attractive potential in a hydrogen atom the term V_2r^2 (with $V_1 = 0$) gives a good representation of the quadratic Zeeman effect in weak magnetic fields. The Zeeman calculation involves simulating an anisotropic potential of form $x^2 + y^2$ by the simple radial potential kr^2. The radial potential gives easier calculations, but the value of k is state dependent and is worked out by using the algebra of angular momentum theory (Killingbeck 1981c).

Further details are given in connection with ZEEMAN; this is yet another case in which a mathematical analysis turns out to produce computational advantages. SERAT takes advantage of some of the recent work on the relationship between the power series and Hill determinant approaches. It is a dual-purpose program, since it will give either Hill determinant results corresponding to the boundary condition $\psi(\infty) = 0$ or power series results corresponding to the boundary condition $\psi(L) = 0$ for some finite L value.

9.3 The recurrence relation

The fundamental recurrence relation used in SERAT is obtained by taking the wavefunction in the form

$$\psi = r^{l+1} \exp(-\beta r - \gamma r^2/2) \sum_0^\infty A(N) r^N. \tag{9.3}$$

The factor r^{l+1} in (9.3) is chosen to produce a radial function for a state with angular momentum l, as the reader may verify by acting on it with the first two terms in the Schrödinger equation (9.1). Inserting (9.3) in that Schrödinger equation gives, after some tedious algebra, the recurrence relation

$$\begin{aligned}(N+2)(N+2l+3)A(N+2) &= [(2N+2l+4)\beta - Z]A(N+1) \\ &\quad + [(2N+2l+3)\gamma - \beta^2 - E]A(N) \\ &\quad + (V_1 - 2\beta\gamma)A(N-1) + (V_2 - \gamma^2)A(N-2).\end{aligned} \tag{9.4}$$

The basic principle of the power series approach is to sum the series appearing in (9.3) and produce a value for $\psi(r = L)$. Given the starting coefficient value $A(0) = 1$, with all $A(N)$ set at zero for $N < 0$, the recurrence relation (9.4) produces the sequence of $A(N)$ values to be used in (9.3). It is clear that the coefficients $A(N)$, and thus also the calculated $\psi(L)$, will depend on the E value used in (9.4). The approach adopted is to vary the trial E value to make $\psi(L)$ zero, corresponding to the use of homogeneous Dirichlet boundary conditions at $r = 0$ and $r = L$. There is an obvious simplification which is possible: since the factors outside the summation in (9.3) are positive and non-zero, it suffices to use the sum of the series to represent $\psi(L, E)$, provided that L is *kept fixed* as E is varied to locate the eigenvalues where $\psi(L, E)$ vanishes. It is clear that the approach described above is well suited to the root-finding methods which are used in many of the programs in this book. To achieve the most simple and flexible version of the power series approach, however, we must look more closely at the sum which is used to represent $\psi(L, E)$.

9.4 The zero-coefficient test

Until recent times the universal technique used to apply the series method was to perform the sum of the terms $A(N)r^N$ until convergence to some numerical value was obtained. Some of the early calculations did not even bother with an exponential factor (i.e. they had $\beta = \gamma = 0$), but it was soon discovered that the use of suitably chosen β (even with $\gamma = 0$) produces more rapid convergence of the series. Ginsberg (1982) noted that a remarkable simplification was possible: he obtained correct energy levels for several potentials by applying the simple test $A(N, E) = 0$, without bothering to sum the series for the wavefunction. The energy levels obtained depended on N, but quickly attained asymptotic values as N was increased. Killingbeck (1985d) showed that Ginsberg's zero-coefficient method is equivalent to the Hill determinant method. In that method the $EA(N)$ term in equation (9.4) is separated out so that the equation becomes a matrix eigenvalue problem. The resulting matrix is clearly a sparse one, with many of the matrix elements being zero. As the truncation of the matrix is carried out at larger and larger N values, the low eigenvalues of the matrix stabilize to give the energy levels of the Schrödinger equation (9.1). Ginsberg's method is simply a way to obtain the matrix eigenvalues by using the recurrence relation directly. It is essentially a kind of shooting calculation, akin to the ones used in the programs RADIAL and FIDIF of this book, except that there is no marked instability involved.

The developments described above still leave one further link to be established: the traditional power series method literally constructs the wavefunction at a point in space, whereas the Hill determinant method uses the more abstract notion of matrix eigenvalues. The link between the two viewpoints is clearly similar to that used to relate wavefunction and matrix concepts in the Rayleigh–Ritz method. If the wavefunction $\psi(x)$ is written in the form

$$\psi(r) = \sum A(N)\phi_N(r) \tag{9.5}$$

for some basis set of functions ϕ_N, then the condition that $\psi(r)$ shall obey the Schrödinger equation leads to a matrix eigenvalue problem with the $A(N)$ being the elements in the eigencolumn. This is the situation which we have here, except that the basis functions used in our expansion (9.3) are certainly not orthonormal. The resulting matrix which leads to the Hill determinant thus differs from the matrix which would result if the proper inner products were formed to produce a Rayleigh–Ritz matrix eigenvalue problem. That Rayleigh–Ritz approach would give a symmetric generalized eigenvalue problem. The Hill determinant approach, on the other hand, is based on an ordinary eigenvalue problem for a non-symmetric matrix. That matrix has complex eigenvalues as well as real ones, but as the dimension is increased the number of real eigenvalues increases and the lower ones converge to the eigenvalues of the Schrödinger equation. That this is so is often established by computational experiment. Hautot (1982) has described a perturbed oscillator problem for which the approach does not work unless the simple power terms used in the series are replaced by Hermite polynomials. The most exhaustive algebraic investigation of the region of validity of the simple Hill determinant approach is contained in the works of Znojil (1976, 1981).

9.5 Forwards nested multiplication

The series to be summed in working out $\psi(L, E)$ is of the form

$$S(M, L) = \sum_{0}^{M} A(N)L^N. \tag{9.6}$$

Since we work out the coefficients $A(N)$ in order of increasing N, it is natural to look for the effect of adding one more term to the sum. The traditional form of nested multiplication for evaluating $S(M, L)$ would start 'from the top'. For $M = 2$, for example, we would have

$$S(2, L) = [A(2)L + A(1)]L + A(0). \tag{9.7}$$

To find $S(3, L)$ we have to go back and start all over again. However, if we introduce the variable $Y = L^{-1}$ we find

$$L^{-2}S(2, L) = [A(0)Y + A(1)]Y + A(2) \tag{9.8}$$

and have a procedure which operates forwards, in the usual direction of summation. There is still a factor L^{-M} involved, of course, so that it is actually $L^{-M}S(M, L)$ which is being evaluated. If the summation for all the trial E values is carried out up to the *same* fixed M value, then the factor L^{-M} can be discarded along with the other factors in equation (9.3). The conclusion of this reasoning is that the wavefunction $\psi(L)$ for a given M can be represented by the sum obtained from the forwards nested multiplication. This sum obeys the recurrence relation

$$S(M + 1) = YS(M) + A(M + 1) \tag{9.9}$$

with $Y = L^{-1}$, where L is the value of r for which we are imposing the boundary condition $\psi(L) = 0$. If we wish to set the boundary at infinity (where it usually is in quantum mechanics textbooks) we simply set $Y = 0$. The eigenvalue test then becomes $S(M + 1) = A(M + 1) = 0$, for some M value; this is just Ginsberg's zero-coefficient test, which is equivalent to the Hill determinant approach! This beautiful result makes it possible to have a one-line statement in a program which can be used to give either a power series result (for finite L) or a Hill determinant result (for $L \to \infty$). This means that the dual-purpose program given here could have been called HILL, to make its features clear to a user.

The name SERAT was chosen on the grounds that the Hill determinant method (at least as derived here) emerges as a special case of the power series method.

9.6 The perturbed oscillator potential

SEROSC employs the same principles as SERAT, but is adapted for use with the perturbed oscillator Schrödinger equation

$$-D^2\psi + [l(l+1)r^{-2} + V_1 r^2 + V_2 r^4 + V_3 r^6]\psi = E\psi. \quad (9.10)$$

Because only even powers of r appear in (9.10) the wavefunction can be taken in the compact form

$$\psi = r^{l+1} \exp(-\beta r^2/2 - \gamma r^4/4) \sum_0 A(N) r^{2N} \quad (9.11)$$

whereas both even and odd powers of r were present in the equations for the perturbed Coulomb problem. Substituting (9.11) into (9.10) leads to the recurrence relation

$$\begin{aligned}(2N + l + 3)(2N + l + 2)A(N+1) &= [(4N + 2l + 3)\beta - E]A(N) + [(4N + 2l + 1)\gamma \\ &\quad + V_1 - \beta^2]A(N-1) + (V_2 - 2\beta\gamma)A(N-2) \\ &\quad + (V_3 - \gamma^2)A(N-3). \quad (9.12)\end{aligned}$$

Most of the comments made in preceding sections about the perturbed Coulomb problem are equally valid for the perturbed oscillator problem, except that the recurrence relation (9.9) becomes

$$S(M+1) = Y^2 S(M) + A(M+1) \quad (9.13)$$

since r^2 rather than r is the relevant variable.

9.7 The 'false eigenvalue' effect

By studying the recurrence relation (9.4) it seems that an appropriate choice of β and γ will make the last two terms vanish, so that only a three-term recurrence relation (or an equivalent tridiagonal matrix eigenvalue problem) will remain. This tempting possibility turns out to be a trap, since it leads to what appear to be obviously incorrect energy levels for the potential of equation (9.1), even though the convergence to those levels is normal as the number of terms taken is increased. However, if γ is held at the value $V_2^{1/2}$, to eliminate the last term in

(9.4), but β is varied, then it is found by computational experiment that there is a critical β value which separates the 'true' and 'false' eigenvalue regimes; the 'natural' β value which produces maximum simplicity in the recurrence relation is (unfortunately) in the 'false' region. A similar effect is found to hold for SEROSC; thus, while it *is* possible to make the automatic choice $\gamma = V_3^{1/2}$ to eliminate *one* term from the recurrence relation, it is still important to keep β adjustable. The work of Znojil (1986) and Killingbeck (1987e) showed that the 'false' energies are not meaningless. They actually correspond to the energy levels of a partner potential in which the signs of some of the potential coefficients are reversed; Znojil (1986) showed this by a study of the symmetry properties of the recurrence relation for the $A(N)$.

SERAT and SEROSC. Programming notes

9.8 Array dimensions

The variables $A(N)$ and $S(N)$ could be stored in arrays, but this has not been done in the programs given here, so that the programs will work even on a microcomputer with limited array capabilities. Only five $A(N)$ at a time are used in the recurrence relations (9.4) and (9.12). This means that only five variables A1 to A5 are needed to keep track of the calculation, with the latest A1 value being the one used to modify S in equations (9.9) or (9.13). S can be a single variable, since only its final value is needed.

9.9 Modular structure

The programs SERAT and SEROSC have the modules ROOTSCAN and SECANT in common. These modules constitute about half of each program while the function subroutine and precomputation module have the same structure in both programs, differing only in the detail of the algebraic expressions appearing in the program lines.

9.10 Precomputation. Reduction in strength

The two programs have been used to illustrate the technique of 'reduction in strength' discussed by Cocke and Kennedy (1977). The idea is to avoid repeated and redundant calculations of the same quantity within a loop and to reduce multiplications to additions, exponentiation to multiplication, etc, on the grounds that this reduces the computing time required. The example of the coefficients of

$A(N+2)$ in equation (9.4) can be used to show how the method is applied. The product $(N+2)(N+2l+3)$ is required for N values from -1 up to some upper limit NT. A little algebra shows that at $N=-1$ the product has the value $2l+2$ and that on passing from N to $N+1$ it increases its value by an amount $2N+2l+6$, while the quantity $2N+2l+6$ increases by two. Thus the value of the product $P=(N+2)(N+2l+3)$ for successive N values can be produced by the sequence of assignment statements

$$P := P + I : I := I + 2 \tag{9.14}$$

at the end of the loop, with the values $P=2l+2$ and $I=2l+4$ being set outside the loop (if $N=-1$ is the intended initial value). This approach can be used for all the coefficients in the recurrence relations (9.4) and (9.12). If the coefficients $A(N)$ are to be named A1 to A5 in the program, then a natural notation is to use the symbols C1 to C5 for the coefficients which multiply them. These coefficients then start off with initial values C1, C2, etc at the initial N value and suffer increments I1, I2, etc on each loop traversal. This simple style and notation (C for coefficient, I for increment) is used in both SERAT and SEROSC and helps to make the structure of the programs easy to follow.

9.11 The scaling factor

If 50 or so terms $A(N)$ are used in the calculation, it may happen that the later $A(N)$ are so diminished as to be below the computer underflow level. This means that for *any* E value the function value will be returned as zero, leading to a ratio 0/0 in the root-finding modules and rendering the root-finding process inoperable. To amplify the function value and keep it above the underflow value a scaling factor K is introduced so that the $A(N)$ are expressed in the form

$$A(N) = B(N)K^{-N} \tag{9.15}$$

with K typically being 1, 2, 4 or 8. The value of $A(0)=B(0)$ can also be increased to 1E20 (for example) instead of one. When the substitution (9.15) is used, the only effect is to multiply the coefficients in the recurrence relations (9.4) and (9.12) by appropriate powers of K. It turns out that the particular array-free style adopted will also allow an approach which incorporates K without modifying the recurrence relation.

9.12 SERAT. Program analysis and program

Lines 10 to 30 are the input lines which specify the potential, the angular momentum and the parameters β and γ (BE and GA). The

K value in line 25 can be changed by using the line editor, or an input line can be substituted. The remmed line 30 gives one possible way of choosing γ automatically.

Lines 40 to 70 involve the setting of the initial values of the coefficients C1, C2, C4 and C5 and of the increments I1, I2 and I3 which are to be added to them as explained in the programming notes. Since C1, C2 an I1 later change value within the loop (in subroutine 1000) they have the extra digit 0 attached to them to signify that what is worked out here is the starting value. C1, C2 and I1 have to be reset to their starting values after each loop, whereas C4, C5 and I3 remain unchanged throughout the program.

Lines 100 and 110 set the reciprocal coordinate $\mathrm{Y} = \mathrm{R}^{-1}$ to be used in the recurrence relation derived in the discussion of §9.5. Y is multiplied by the scaling factor.

Lines 200 and 210 are input lines for the number of terms of the series to be used and for the E0 and DE values required by the root-finding process.

Lines 300 to 490 are the two modules which carry out the root-finding process. SECANT could be replaced by one of the other modules such as NEWTON. The control on the shift size in line 470 has been remmed, but can be activated if necessary. The convergence threshold in line 460 can be adjusted as required; the PRINT statement is not essential, but produces a helpful space in the display after a root has been found.

Lines 1000 to 1130 constitute the function subroutine.

Line 1000 sets the coefficient C3, which depends on E, by subtracting E from the precomputed C30.

Line 1010 sets the initial $A(N)$ values and sets the sum of the series to $A(0)$. The symbol W (for wavefunction) is used to denote the quantity $S(N)$ which was discussed in §9.5. A2 is K times the $A(0)$ value, while A3 to A5 represent the $A(N)$ with negative N.

Line 1020 restores the starting values of those coefficients CN which change their value during the operations within the loop.

Line 1030 could have any starting index, since N is only a dummy counting index in the program as presented here. The coefficients C1 and C2 *are* functions of N, as the appropriate mathematical theory showed. In a 'head-on' approach, for example, C1 would be worked out as the product $(N+2)(N+2l+3)$ and so it would be important to start at $N=-1$ to work out $A(1)$. However, in the 'reduction in strength' approach it is the initial precomputed coefficients which set the N value at -1, with N not appearing *explicitly* in the expressions contained within the loop.

Lines 1040 and 1050 work out the current A1, which is the $A(N+2)$ of the original recurrence relation (9.4). The C5*A5 term is separated out because it can often be eliminated (see §9.7).

Lines 1060 and 1070 revise the coefficients C1, C2 and C3 ready for the next loop, in which N changes to N + 1. This step is explained in §9.10.

Lines 1080 and 1090 move along the values of the AN so that A1 is always the one about to be calculated, that is the one on the left of the recurrence relation (9.4). The scaling factor K of §9.11 appears here; it is clear that it is actually introducing factors of K^N into the particular terms in the recurrence relation, but it does so by means of the A factors and not the C factors.

Line 1100 applies the simple recurrence relation derived in §9.5. The only modification required should be the use of a factor K to allow for the scaling which has been applied in the previous line. However, this factor is already included in the Y calculation of line 110.

Line 1120 sets the function value returned by the subroutine equal to the W value; this is equivalent to imposing the boundary condition $\psi(R) = 0$ for the wavefunction, provided that NT is sufficiently large.

The structure of the program SEROSC is the same, except that the detailed form of the C and I parameters is adapted to the specific recurrence relation involved. Y^2 is multiplied by K, since it is the quantity involved in the power series.

```
  7 REM  **********************************
  8 REM  SERAT
  9 REM  **********************************
 10 INPUT "Z,V1,V2";Z,V1,V2
 15 INPUT "ANG.MOM";L
 20 INPUT "BETA,GAMMA";BE,GA
 25 LET K=2
 30 REM LET GA=SQR V2
 37 REM  **********************************
 38 REM  PRECOMPUTE C AND I
 39 REM  **********************************
 40 LET C10=2*L+2
 45 LET C20=(2*L+2)*BE-Z
 50 LET C30=(2*L+1)*GA-BE*BE
 55 LET C4=V1-2*BE*GA
 60 LET C5=V2-GA*GA
 65 LET I10=2*L+2
 70 LET I2=2*BE: LET I3=2*GA
 91 REM  **********************************
100 INPUT "RADIUS R";R
110 LET Y=K/R: LET Y2=Y*Y
199 REM  **********************************
200 INPUT "NO.OF TERMS";NT
210 INPUT "E,DE";E0,DE
```

```
 297 REM  ********************************
 298 REM  ROOTSCAN
 299 REM  ********************************
 300 LET E=E0
 310 GO SUB 1000: PRINT E,F
 320 LET F1=F: LET E1=E
 330 LET E=E+DE
 340 GO SUB 1000: PRINT E,F
 350 LET F2=F: LET E2=E
 360 LET R=F2/F1
 370 REM IF R>1 THEN LET DE=-DE/4: GO TO 310
 380 IF R>0 THEN GO TO 320
 381 REM  ********************************
 397 REM  ********************************
 398 REM  SECANT
 399 REM  ********************************
 400 LET E=E2-DE+DE/(1-R)
 410 LET LIM=ABS (DE/4)
 420 LET FS=F2: LET ES=E2
 430 GO SUB 1000: PRINT E,F
 440 LET SH=F*(E-ES)/(FS-F)
 450 LET ES=E: LET FS=F
 460 IF ABS (SH/E)<2E-8 THEN PRINT: GO TO 490
 470 REM IF ABS SH>LIM THEN LET SH=LIM*SGN SH
 480 LET E=E+SH: GO TO 430
 487 REM  ********************************
 488 REM  RETURN TO ROOTSCAN
 489 REM  ********************************
 490 LET E=E2: GO TO 310
 491 REM  ********************************
 997 REM  ********************************
 998 REM  HILL ATOM
 999 REM  ********************************
1000 LET C3=C30-E
1010 LET A3=0: LET A4=0: LET A5=0: LET W=1E10: LET A2=W*K
1020 LET C1=C10: LET C2=C20: LET I1=I10
1030 FOR N=0 TO NT
1040 LET S=C2*A2+C3*A3+C4*A4
1050 LET A1=(S+C5*A5)/C1
1060 LET I1=I1+2: LET C1=C1+I1
1070 LET C2=C2+I2: LET C3=C3+I3
1080 LET A5=A4*K: LET A4=A3*K
1090 LET A3=A2*K: LET A2=A1*K
1100 LET W=W*Y+A1
1110 NEXT N
1120 LET F=W
1130 RETURN
1131 REM  ********************************
```

```
  7 REM  ********************************
  8 REM  SEROSC
  9 REM  ********************************
 10 INPUT "V1,V2,V3";V1,V2,V3
 15 INPUT "ANG.MOM";L
 20 INPUT "BETA,GAMMA";BE,GA
 25 LET K=4
 30 REM LET GA=SQR V3
 37 REM  ********************************
 38 REM  PRECOMPUTE C AND I
 39 REM  ********************************
 40 LET C10=L*(L+5)+6
 45 LET C20=(2*L+3)*BE
 50 LET C30=(2*L+1)*GA+V1-BE*BE
 55 LET C4=V2-2*BE*GA
 60 LET C5=V3-GA*GA
 65 LET I10=4*L+6: LET I2=4*BE
 70 LET I3=4*GA
 91 REM  ********************************
100 INPUT "RADIUS R";R
110 LET Y=K/R: LET Y2=Y*Y
199 REM  ********************************
200 INPUT "NO.OF TERMS";NT
210 INPUT "E,DE";E0,DE
297 REM  ********************************
298 REM  ROOTSCAN
299 REM  ********************************
300 LET E=E0
310 GO SUB 1000: PRINT E,F
320 LET F1=F: LET E1=E
330 LET E=E+DE
340 GO SUB 1000: PRINT E,F
350 LET F2=F: LET E2=E
360 LET R=F2/F1
370 REM IF R>1 THEN LET DE=-DE/4: GO TO 310
380 IF R>0 THEN GO TO 320
381 REM  ********************************
397 REM  ********************************
398 REM  SECANT
399 REM  ********************************
400 LET E=E2-DE+DE/(1-R)
410 LET LIM=ABS (DE/4)
420 LET FS=F2: LET ES=E2
430 GO SUB 1000: PRINT E,F
440 LET SH=F*(E-ES)/(FS-F)
450 LET ES=E: LET FS=F
460 IF ABS (SH/E)<2E-8 THEN PRINT: GO TO 490
470 REM IF ABS SH>LIM THEN LET SH=LIM*SGN SH
```

```
 480 LET E=E+SH: GO TO 430
 487 REM  *********************************
 488 REM  RETURN TO ROOTSCAN
 489 REM  *********************************
 490 LET E=E2: GO TO 310
 491 REM  *********************************
 997 REM  *********************************
 998 REM  HILL OSC.
 999 REM  *********************************
1000 LET C2=C20-E
1010 LET A3=0: LET A4=0: LET A5=0: LET W=1E10: LET A2=K*W
1020 LET C1=C10: LET C3=C30: LET I1=I10
1030 FOR N=0 TO NT
1040 LET S=C2*A2+C3*A3+C4*A4
1050 LET A1=(S+C5*A5)/C1
1060 LET I1=I1+8: LET C1=C1+I1
1070 LET C2=C2+I2: LET C3=C3+I3
1080 LET A5=A4*K: LET A4=A3*K
1090 LET A3=A2*K: LET A2=A1*K
1100 LET W=W*Y2+A1
1110 NEXT N
1120 LET F=W
1130 RETURN
1131 REM  *********************************
```

In most applications the author has found it adequate to set γ at the value given by activating line 30. The γ input in line 20 then does not actually control the γ value. The results of tables 9.1 and 9.2 below were obtained with this choice of γ.

SERAT and SEROSC. Specimen results

9.13 Published applications for SERAT

The Schrödinger equation with a potential of the type $-Zr^{-1} + V_1 r + V_2 r^2$ has been used in conjunction with the power series–Hill approach of SERAT for several published calculations, so that a plentiful supply of numerical results can be found in the literature. The applications include:

(1) a study of the 'false eigenvalue' problem for the perturbed Coulomb potential when treated by the Hill determinant method (Killingbeck 1986b);
(2) a study of the calculation of expectation values of type $\langle r^N \rangle$ by means of the power series method (Killingbeck 1985c);
(3) a study of the rotating displaced oscillator problem, which has the special feature $V_1 = -2V_2$ (Killingbeck 1987d);

(4) the calculation of weak-field energy levels for the hydrogen atom quadratic Zeeman effect (Killingbeck 1979b, 1981c).

9.14 Some specimen results for SEROSC

Tables 9.1 and 9.2 give SEROSC results for a single- and a double-well potential. In table 9.2 the very small energy splitting between the lowest even- and odd-parity states and the much larger splitting for the first excited states, which have positive energy and do not involve wavefunctions concentrated in the potential wells.

Table 9.1 Even-parity energies ($V_1 = V_2 = V_3 = 1$, $\beta = 7$).

CT/NT	30	40	50
0 → 1	1.614 894 1	1.614 894 1	1.614 894 1
1 → 2	11.107 354	11.107 353	11.107 353
2 → 3	25.068 681	25.068 671	25.068 671
3 → 4	42.236 875	42.236 729	42.236 729
4 → 5	62.063 330	62.061 886	62.061 886
5 → 6	84.219 562	84.208 895	84.208 891
6 → 7	108.505 38	108.445 00	108.444 97
7 → 8	132.860 15	134.596 35	134.596 10
8 → 9	163.422 35	162.527 35	162.525 76
9 → 10	194.535 0	192.131 72	192.123 07

Table 9.2 Energies for $V = x^2 - x^4 + 0.1x^6 (\beta = 5)$.

Parity, CT/NT	60	70	80
Even, 0 → 1	−4.314 936 2	−4.315 074 1	−4.315 073 5
Odd, 0 → 1	−4.312 500 2	−4.312 611 0	−4.312 611 8
Even, 1 → 2	0.425 489 1	0.425 394 8	0.425 393 9
Odd, 1 → 2	1.352 960 4	1.352 735 4	1.352 733 4

9.15 The CT parameter

The parameter CT which appears in the tables is a parameter which counts the number of sign changes in the computed $A(N)$. Theoretically, because of the link between the power series and matrix approaches, this number might be expected to increase by one as E passes through an eigenvalue, and so could be used to indicate the particular state concerned. An extra line in the program is easily inserted to compute CT. This was done on an *ad hoc* basis while computing the specimen results, since the results for very high energy levels seem to show some irregularity, and this is still under investigation. For states with state number well below NT, however, the use of CT has been found to be reliable. We conjecture that this is because of the existence of a group of all real eigenvalues at the bottom of the spectrum, the size of the group increasing with NT.

MOMOSC. Mathematical theory

9.16 Inner product recurrence relations

MOMOSC is designed to deal with the Schrödinger equation

$$H\psi = -D^2\psi + \sum_{M=0}^{3} V_{2M}x^{2M}\psi = E\psi \tag{9.16}$$

although its extension to include further potential terms is straightforward. The first step in the mathematical theory is to take the inner product of equation (9.16) from the left with the function $x^N f$, where f is the reference function

$$f = x^P \exp(-\beta x^2/2). \tag{9.17}$$

The parity index P is zero for even states and one for odd states. The left-hand side of the resulting inner product equation

$$\langle f|x^N H|\psi\rangle = E\langle f|x^N|\psi\rangle \tag{9.18}$$

can be evaluated further, either by an integration by parts or by using the rule of Dirac algebra which says that the left-hand side equals $\langle\psi|Hx^N|f\rangle$ if ψ is real. The only term which requires any detailed work is the one involving $\langle\psi|D^2x^N f$, but it is easy to see that it will yield a sum of terms involving the quantities

$$S(N) = \langle f|x^N|\psi\rangle. \tag{9.19}$$

Equation (9.18), with the left-hand side fully evaluated by using the H operator of equation (9.16), yields the recurrence relation

$$\begin{aligned}(N+P)(N+P-1)S(N-2)& \\ = [(2N+2P+1)\beta - E]S(N) &+ (V_2-\beta^2)S(N+2) \\ &+ V_4S(N+4) + V_6S(N+6).\end{aligned} \tag{9.20}$$

A computation based on the $S(N)$ is possible, and would be similar in form to the one used in SEROSC. For variety, we adopt a different approach here. Rather than allowing for the possibility of scaling the $S(N)$ when they produce overflow or underflow, we avoid the scaling problem by working in terms of ratios $R(N)$ which are defined by the equation

$$S(N+2) = R(N)S(N). \tag{9.21}$$

Dividing equation (9.20) by $S(N)$ gives the result

$$R(N-2) = (N+P)(N+P-1)/T(N) \tag{9.22}$$

with

$$T(N) = [(2N+2P+1)\beta - E] + (V_2 - \beta^2)R(N) + V_4R(N)R(N+2) + V_6R(N)R(N+2)R(N+4). \tag{9.23}$$

For the special case $N = 0$ we obtain the result

$$F(E) = 0 \tag{9.24}$$

where the function $F(E)$ is given by

$$F(E) = [(2P+1)\beta - E] + (V_2 - \beta^2)R(0) + V_4R(0)R(2) + V_6R(0)R(2)R(4). \tag{9.25}$$

The derivation of (9.25) involves no assumption about $S(-2)$, since it follows directly on taking the inner product of (9.16) with f.

9.17 The computational procedure

The calculation so far has produced a root-finding problem, and so fits in with the style adopted in many of the programs of this book. It only remains to specify the method of calculation of the function $F(E)$ in more detail. Blankenbecler *et al* (1980) in their application of the method made a study of the asymptotic behaviour of the $S(N)$ at large N, using the results to start off the recurrence relation (9.20) correctly at a large fixed N value N_0. However, it was discovered computationally (Killingbeck *et al* 1985, Killingbeck 1987e) that the simple standard choice $R(N) = 0$ for $N > N_0$ suffices if (9.22) is used. This choice, with E variable, produces definite roots for the function $F(E)$. These roots vary with N_0, but tend to asymptotic values as N_0 is increased. As N_0 is increased the dependence of the roots on the parameter β becomes weaker; a wide plateau in β develops, with the energy levels obtained being independent of β across it. The early calculations using the inner product approach calculated ground-state energies, but application of the equation (9.22) to calculate excited state energies (Killingbeck *et al* 1985) revealed a special property of the function $F(E)$. Each zero of $F(E)$, except the lowest, has a tan-type singularity close to it. The gap between the zero and the singularity decreases as E increases but increases as β increases. This means that the calculation of high excited states requires some care; the DE parameter in the ROOTSCAN module must be sufficiently small compared with the gap between the zero being sought and its associated singularity.

The quantities $S(N)$ of equation (9.19) are inner products; many authors refer to them as moments, although they should not be confused with the diagonal moments $\langle \psi|x^N|\psi\rangle$ which appear in the hypervirial equations of HYPOSC. Vorobyev (1965) and Handy and Williams (1987) use this terminology, and so the program given here has been given the name MOMOSC (moment-oscillator).

9.18 A link with matrix theory

The use of the ratio quantities $R(N)$ in the computation has the important advantage that no scaling of the variables is necessary, but the resulting function $F(E)$ has a singularity structure which means that the calculation of high excited states requires extra care.

Fortunately this care is possible when interactive computing is carried out on a microcomputer, but it is interesting to reflect on the possible source of the problem. A similar phenomenon was noted in connection with the program FOLDER. For the matrix eigenvalue problem treated there the approach via perturbation theory led to a function $F(E)$ with both zeros and poles, whereas the approach finally adopted removed the poles. In the present case the use of the ratio quantities $R(N)$ gives a function with zeros and poles; the use of the $S(N)$ directly, although it requires care about scaling difficulties, does not lead to difficulty with poles. A study of the method of computation used here makes it clear that it is essentially a special way of finding the eigenvalues of a matrix. Equation (9.20) gives the matrix elements and it is clear that the matrix is of upper Hessenberg type, with only one subdiagonal having non-zero elements. Such a matrix can have its eigenvalues located quite easily by recurrence relation methods. Killingbeck (1987e) discussed some of the links between the power series, inner product and matrix diagonalization techniques, showing their equivalence. During the writing of this book it has become increasingly clear to the present author how, despite the varying mathematical concepts from which they start, many of the methods of eigenvalue calculation in quantum mechanics are disguised calculations of eigenvalues for matrices which have special simple forms, the Hessenberg form being the most prevalent. One of the standard procedures of orthodox matrix theory is to reduce a matrix to tridiagonal form, on the grounds that the eigenvalues of a tridiagonal matrix can be found easily by recurrence relation techniques (Wilkinson 1965). However, the eigenvalues of a Hessenberg matrix can be found almost as easily using recurrence relations. The inner product and other techniques are ways of calculating those eigenvalues, for the special case in which the matrix elements are given by an algebraic formula rather than being given fixed numerical values. Increasing N_0 is obviously increasing the number of basis functions, while varying β is a crude way of varying collectively the whole set of basis functions used. The recurrence relation approach makes it quite feasible to handle what are in effect matrices of large dimension (e.g. 100×100), which would require a great amount of storage space in an approach such as FOLDER which explicitly stores and transforms numerical matrix elements.

9.19 The angular momentum variable

MOMOSC as presented here is a transcription into modular form of a program written during work on a scientific paper (Killingbeck *et al* 1985). The original work dealt only with the perturbed oscillator in one dimension, so that the parameter P became a parity indicator. However MOMOSC is of more general applicability. Repeating the mathematical analysis of §9.16 for the radial Schrödinger equation shows that the use of the reference function (9.17) with $P = l + 1$ is appropriate for states of angular momentum l. Further (Killingbeck 1985d), a change of variable for the radial Schrödinger equation in N dimensions can be used to make it look like the Schrödinger equation in three dimensions. The relationship between the generalized angular momentum Λ in N dimensions and the resulting effective angular momentum l in the three-dimensional reduced version is

$$l = (2\Lambda + N - 3)/2. \tag{9.26}$$

The consequence of this algebra for the program MOMOSC is very simple: P is set equal to $l + 1$, with l given by (9.26). For example, to treat a fully isotropic ($\Lambda = 0$) state in two dimensions, we set $P = 1/2$. For a fully isotropic state in one dimension (i.e. an even-parity state) we set $P = 0$. The case $P = 1/2$ has been tested on MOMOSC for the case $V_2 = 1$, $V_4 = 0.1$, since accurate results for the two-dimensional perturbed oscillator (obtained by another method) have been published by Killingbeck and Jones (1986). The results given by MOMOSC were fully in accordance with the published results.

9.20 A perturbation approach

If the inner products $S(N)$ and the energy E are formally expanded as a power series in λ, with the potential taken as $V_2x^2 + \lambda V_4x^4$ (for example), then a set of recurrence relations results which permits a computation of the perturbed energy (Killingbeck *et al* 1985). The parameter β plays a role akin to that of the K parameter in the hypervirial method (HYPOSC). However, HYPOSC is superior in that it gives the $\langle x^N \rangle$ and also the WKB results. For the two-dimensional perturbed oscillator, however, the inner product perturbation theory works (Killingbeck and Jones 1986), while the hypervirial method does not seem capable of application.

MOMOSC. Programming notes

9.21 The singularity problem

In the main there are no new principles involved in MOMOSC, since it fits into the root-finding approach and simply requires the construction of the appropriate function module to calculate the function $F(E)$ of equation (9.25). However, that function has tan-type singularities at some E values. Such a singularity will give a sign change in ROOTSCAN, which will then transfer control to SECANT. What is required is a filter at the start of SECANT which will send control back to ROOTSCAN if the 'zero' is a bogus one. The interpolated E value, based on the function values at E1 and E2, is what allows us to construct this filter. Studying the graph of a tan-type singularity makes it clear that, if E1 and E2 closely straddle the singularity, then $F(E)$ must exceed in absolute value at least one of F(E1) and F(E2). For a zero this will not be the case. Thus a test on $F(E)$ can be performed to reject the singularities; this makes clear once again the wisdom of storing within ROOTSCAN the values of F(E1) and F(E2).

9.22 MOMOSC. Program analysis and program

Lines 10 to 60 set up the specification of the potential, the parity, the number of terms to be used (called NT rather than N0) and the parameter β which modifies the basis functions. The name A(N) rather than R(N) is used.

Line 200, for the input of the E0 and DE values for the root finder, is given a simple line number, so that the manual instruction GO TO 200 can be used to vary the region and fineness of the search interactively. For higher states DE will have to be smaller than for lower states.

Lines 300 to 490 are the usual pair of root-finding modules. The only novelty is in the extra interpolated line, line 435. This applies the idea explained in the programming notes, returning the root scan to E2 if the proposed 'root' is a singularity. DE must be sufficiently small, of course, for the zero and its associated singularity to be seen separately. If they both fall within the interval DE then no sign change will be detected and the scan will continue.

Lines 1000 to 1090 calculate the function $F(E)$.

Line 1000 zeros the starting A(N). This is not important if NT is being gradually *increased*, since higher A(N) will have been initialized at zero when the run started.

Line 1010 (with line 1050) apply a little of the 'reduction in strength' idea (used in SEROSC) to the coefficient $(2N+1+2P)\beta - E$ which appears in the recurrence relation. This could also be done for the coefficient in the line 1040, but the program operates quite speedily without this further step.

Line 1030 A nested multiplication approach to the sum of products saves one multiplication in working out the sum of products required. The same approach is used in line 1070.

Line 1080 sets the function $F(E)$ for return to the calling module. If the positions of the singularities are required (rather than those of the zeros) the extra statement LET F = 1/F will suffice to reverse the roles of the zeros and singularities.

The subroutine which evaluates $F(E)$ uses a step of -2 in the loop starting at line 1000; it is clear that only about half of the R(N) are used while the rest simply retain a zero value. Most modern microcomputers can cope with an array of a few hundred elements, so the inefficient packing of information into the R array is not harmful. Indeed, if odd-parity potential terms such as $V_1 x$ and $V_3 x^3$ are introduced into the problem then *all* the R(N) will be used, although the mathematical theory will need appropriate modification. Killingbeck (1988d) gave an account of how the method of MOMOSC can be modified to deal with a potential in which a small term of x^3 type perturbs an initial potential with x^2 and x terms. For an even-parity potential it would be possible roughly to halve the required A array size by relabelling the elements so that the step in line 1000 becomes -1 instead of -2. Alternatively, the use of an array could be avoided by using the approach explained for the programs SERAT and SEROSC.

```
 7 REM  **********************************
 8 REM  MOMOSC
 9 REM  **********************************
10 INPUT "V2,V4,V6";V2,V4,V6
20 INPUT "PARITY,0 OR 1";P
30 INPUT "NT (EVEN)";NT
```

```
  40 DIM A(NT+5): LET PP=P+P
  50 INPUT "BETA";B
  60 LET U2=V2-B*B
 199 REM  *********************************
 200 INPUT "E0,DE";E0,DE
 297 REM  *********************************
 298 REM  ROOTSCAN
 299 REM  *********************************
 300 LET E=E0
 310 GO SUB 1000: PRINT E,F
 320 LET F1=F: LET E1=E
 330 LET E=E+DE
 340 GO SUB 1000: PRINT E,F
 350 LET F2=F: LET E2=E
 360 LET R=F2/F1
 370 REM IF R>1 THEN LET DE=-DE/4: GO TO 310
 380 IF R>0 THEN GO TO 320
 381 REM  *********************************
 397 REM  *********************************
 398 REM  SECANT
 399 REM  *********************************
 400 LET E=E2-DE+DE/(1-R)
 410 LET LIM=ABS (DE/4)
 420 LET FS=F2: LET ES=E2
 430 GO SUB 1000: PRINT E
 435 IF ABS F>ABS F1 OR ABS F>ABS F2 THEN LET E=E2: GO TO 310
 440 LET SH=F*(E-ES)/(FS-F)
 450 LET ES=E: LET FS=F
 460 IF ABS (SH/E)<2E-8 THEN PRINT: GO TO 490
 470 REM IF ABS SH>LIM THEN LET SH=LIM*SGN SH
 480 LET E=E+SH: GO TO 430
 487 REM  *********************************
 488 REM  RETURN TO ROOTSCAN
 489 REM  *********************************
 490 LET E=E2: GO TO 310
 997 REM  *********************************
 998 REM  MOMENT CALC.
 999 REM  *********************************
1000 LET A(NT+1)=0: LET A(NT+3)=0: LET A(NT+5)=0
1010 LET C=(NT+NT+1+PP)*B-E: LET IC=-4*B
1020 FOR N=NT TO 2 STEP -2
1030 LET T=C+((A(N+5)*V6+V4)*A(N+3)+U2)*A(N+1)
1040 LET A(N-1)=N*(N+PP-1)/T
1050 LET C=C+IC
1060 NEXT N
1070 LET S=((A(5)*V6+V4)*A(3)+U2)*A(1)
1080 LET F=S+B*(1+PP)-E
1090 RETURN
1091 REM  *********************************
```

MOMOSC. Specimen results

9.23 Comparison with SEROSC

Killingbeck (1987e) gave results to show how the calculated energy varies with NT and β for the special case of the sixth even-parity state for the potential $x^2 + x^4$.

Table 9.3 shows some results for the potential $x^2 + x^4 + x^6$, with β held at the value 10 and NT made sufficiently large to give convergence of the particular eigenvalue quoted. For the higher energy levels the value of the search interval DE has to be made small. This arises because each energy level except the lowest has near to it a tan-like singularity. This singularity will give a sign change in $F(E)$ if DE is very small and thus trigger the use of the secant method root finder, which will move to the nearest zero of $F(E)$. However, if DE is too large, the interval of length DE may contain both the singularity and the root; the program then detects no sign change and misses the root, since the two sign changes (one from the singularity, one from the zero) both occur within the interval. Table 9.3 shows the location of the singularity associated with each zero of $F(E)$; for example, the notation 108.3–4 means that the singularity is in the interval between 108.3 and 108.4

The energy levels displayed in table 9.3 were also given in the specimen results for the program SEROSC; this makes it clear that the two methods agree, although SEROSC does not give the problem with singularities which arises for excited states in MOMOSC. As a further test of MOMOSC we calculated the lowest energy levels for the potential $x^2 - x^4 + 0.1x^6$, with $\beta = 5$. To obtain convergence an NT of 200 was needed; a DE value of 0.01 was needed, since for this double-well potential it is found that even the ground state of each parity has a

Table 9.3 Even-parity results for $V = x^2 + x^4 + x^6$.

E	(NT, DE)	Singularity
1.614 894 1	100, 0.5	
11.107 353	100,0.5	7.8–9
25.068 671	100,0.5	22.6–7
42.236 729	100,0.5	40.7–8
62.061 886	100,0.5	61.3–4
84.208 891	100,0.5	83.8–9
108.444 97	120,0.1	108.3–4
134.596 10	140,0.01	134.55–6
162.525 75	140,0.01	162.51–2
192.123 07	160,0.001	192.11–2

partner singularity, at a slightly *higher* energy than the zero. The even- and odd-parity ground-state energies were found to be at −4.315 075 2 and −4.312 611 8, respectively, agreeing with the SEROSC specimen results.

9.24 Some specimen expectation values

Adding the small perturbing term εx^N to the potential in the Schrödinger equation changes each energy level by an amount $\varepsilon\langle x^N\rangle$ (to first order in ε), where $\langle x^N\rangle$ is the expectation value of x^N with respect to the eigenfunction with energy E. A central difference estimate of $\langle x^N\rangle$ can thus be found using the formula

$$2\varepsilon\langle x^N\rangle = E(V + \varepsilon x^N) - E(V - \varepsilon x^N). \tag{9.27}$$

Table 9.4 shows some results for the second even-parity eigenstate of the potential $x^2 + x^4$, showing $\langle x^2\rangle$ and $\langle x^4\rangle$ as calculated from energy differences, plus Richardson extrapolation where necessary. The hypervirial relations which are used in the perturbation theory program HYPOSC can be used here to relate the values of the $\langle x^N\rangle$ for different N. In particular, the lowest-order hypervirial theorem is the traditional virial theorem, which gives the result $E = 2\langle x^2\rangle + 3\langle x^4\rangle$ for the case of the potential $x^2 + x^4$. The numerically calculated values for the three quantities correctly obey this theoretical relation.

Table 9.4 Results for $V_2x^2 + V_4x^2$, with $\beta = 10$, NT = 100.

	V_2	V_4	E
1	1	1.02	8.697 108 6
2	1	1.01	8.676 138 0
3	1	0.99	8.633 842 9
4	1	0.98	8.612 515 1
5	1.02	1	8.678 141 9
6	1.01	1	8.666 600 2
7	0.99	1	8.643 491 4
8	0.98	1	8.631 924 2
	1	1	8.655 050 0

$\langle x^4\rangle = 2.114\,84$ from 1 and 4
$= 2.114\,75$ from 2 and 3
$= 2.114\,72$, using Richardson extrapolation
$\langle x^2\rangle = 1.155\,44$ from 5 and 8
$= 1.155\,44$ from 6 and 7

10 Two research problems

Programs

TWODOSC, ZEEMAN.

10.1 General introduction

The two programs described in this chapter arose in the author's recent research, and are not yet as fully 'user-friendly' as they could be. Nevertheless, they perform their intended tasks accurately; the reader is warned about the correct sequence of operations to carry out in using them, and some readers will perhaps have the patience to make the programs completely foolproof. TWODOSC applies a combination of the recurrence relation method and the iterative method of HITTER to what is essentially a large matrix eigenvalue problem, namely the calculation of the energy levels of a perturbed oscillator in two dimensions. ZEEMAN treats the hydrogen atom Zeeman effect by using a novel combination of shooting and relaxation techniques. The two calculations illustrate how far the simple methods of this book can be taken. Indeed, ZEEMAN produces the most accurate published eigenvalues yet for the hydrogen atom Zeeman effect, and is a remarkable example of how even a small microcomputer can be used to advantage when the underlying mathematics behind the algorithms is formulated with care.

TWODOSC. Mathematical Theory

10.2 The recurrence relations

TWODOSC is a program which applies to a two-dimensional perturbed oscillator the same kind of mathematical method used in the Hill determinant approach to the one-dimensional perturbed oscillator (in the program SEROSC). The Schrödinger equation is taken to have the form

$$-\nabla^2\psi + V_2(x^2 + y^2)\psi + \lambda x^2 y^2 \psi = E\psi \tag{10.1}$$

which represents the most simple case of an oscillator perturbed by a non-separable potential of even parity. The high symmetry of the equation makes it possible to cut down the amount of computation needed in the program. Mathur and Harmony (1977) discussed the C_{4v} point-group symmetry of this Schrödinger equation, but their method of numerical calculation was that of matrix diagonalization using a basis set of unperturbed oscillator product functions. More recently, Fernandez *et al* (1985) have performed accurate eigenvalue calculations for this problem using properly scaled oscillator product functions. Their method of eigenvalue calculation was the iterative one which has been applied in the program HITTER of this book. TWODOSC is a novel hybrid, since it uses that iterative approach (modified to allow for symmetry) to calculate the eigenvalues, but uses a non-orthogonal basis for which the matrix elements arise directly from some simple algebra.

The algebraic manipulations needed to derive the required results are similar to those which have been used previously in connection with the one-dimensional Schrödinger equation. The wavefunction is taken to have the form

$$\psi = \exp[-\tfrac{1}{2}\beta(x^2 + y^2)] \sum A(M, N)x^M y^N. \tag{10.2}$$

When this postulated ψ is substituted in the Schrödinger equation (10.1) it takes a few lines of careful algebra to show that the $A(M, N)$ must obey the recurrence relation

$$[2\beta(M + N + 1) - E]\, A(M, N) = F(M, N) \tag{10.3}$$

where

$$\begin{aligned} F(M, N) = {} & (M + 2)(M + 1)A(M + 2, N) \\ & + (N + 2)(N + 1)A(M, N + 2) \\ & + (\beta^2 - V_2)[A(M - 2, N) + A(M, N - 2)] \\ & - \lambda A(M - 2, N - 2). \end{aligned} \tag{10.4}$$

The modified form which the last term in $F(M, N)$ would take for a symmetric perturbing potential of type $x^A y^B + x^B y^A$ will be clear to the reader, but here we stick to the case $A = B = 2$ throughout.

By rearranging (10.3) and (10.4) it would be possible to produce a matrix eigenvalue problem for which the $A(M, N)$ constitute the eigencolumn. Hajj (1982) adopted this approach, and devised a method which involves making a determinant $D(E)$ equal to zero by varying E. That method is similar in spirit to the one used in the program FOLDER, and would fit into the root-finding approach adopted throughout this book. For the sake of variety, however, we adopt the simple iterative approach to the problem. This proceeds as follows. First, a particular pair of small integers M_0 and N_0 are chosen as being appropriate to pick out some particular state, and $A(M_0, N_0)$ is set equal to one. All the $A(M, N)$ with $(M, N) \neq (M_0, N_0)$ are then adjusted according to the assignment statement

$$A(M, N) := F(M, N)[2\beta(M + N + 1) - E]^{-1} \tag{10.5}$$

for some fixed β and some trial E value. Some cut-off value on M and N is specified, so that higher $A(M, N)$ elements are held at the value zero. After this adjustment process a revised E estimate is calculated using the assignment statements

$$F := 2\beta(M_0 + N_0 + 1) - F(M_0, N_0) \tag{10.6}$$

$$E := RF + (1 - R)E. \tag{10.7}$$

The relaxation parameter R can be decreased in value to help in stabilizing convergence to a desired eigenvalue, as was the case with the program HITTER.

TWODOSC. Programming notes

10.3 Array requirements

A two-dimensional array for the coefficients $A(M, N)$ is obviously needed, with some shifting of indices to avoid M or N being zero for these microcomputers which do not allow zero array indices. To speed up the calculation it is also worthwhile to have two other arrays: the array G stores the products $(M + 1)(M + 2)$ and the array D stores the diagonal elements $2\beta(M + N + 1)$. Both of these arrays are one-dimensional.

10.4 Exploiting the symmetry

For the low-lying singly degenerate states it is possible to assign a parity PX (for the interchange $x \to -x$), a parity PY (for the interchange $y \to -y$) and a parity PXY (for the interchange $x \leftrightarrow y$). As a consequence, some of the $A(M, N)$ may be constrained to be zero while others are related in value. For example, the relationship

$$A(N, M) = \text{PXY}\, A(M, N) \tag{10.8}$$

will hold, and if PXY is -1 it will also follow that the diagonal elements $A(M, M)$ are all zero. This means that only about half of the $A(M, N)$ need to be calculated, since the rest can be filled in by symmetry. The values of M and N used in TWODOSC have intervals of two between them. This is deliberate, for two reasons. First, the even and odd array elements *are* both used (although not simultaneously), depending on the choice of M_0 and N_0. Second, the even and odd $A(M, N)$ must be used together if the program is modified to deal with potential terms such as $xy^3 + x^3y$ which appear in the so-called Henon–Heiles potential. Packing the array down so that every element is used at all times would actually make it less easy to adapt the program to handle such alternative potentials; a similar choice was also made in the case of the program MOMOSC.

10.5 TWODOSC. Program analysis and program

Lines 10 and 20 set up the A, G and D arrays and fill in the elements of G. Fixed large array dimensions have been used in the version shown here, with chosen submatrices being used during the actual computation.

Lines 30 to 45 are input lines to specify which state is being considered. PXY is set at 1 or -1 in line 30. In line 40, however, PX and PY are to be set at the *different* values 0 (for even) or 1 (for odd). M0 and N0 specify which A(M, N) is to be set equal to one, and line 45 ensures that the initial array has the correct $x-y$ symmetry.

Line 50 takes the input V2 and β and precomputes the quantity $\beta^2 - \text{V2}$ which appears in the formulae. With the β value assigned it is now possible to work out the elements of the D array; this is done in line 60.

Line 70 takes the inputs λ, Q and E. Q fixes the maximum number of M and N array indices to be used in the calculation, and E is the initial E estimate, which experience suggests is best set at a value a little higher than the required eigenvalue.

Lines 80 to 130 perform one cycle of adjustment of the A(M, N), with the required function F(M, N) being obtained by a call to the subroutine at line 200. Line 120 performs this call, and fills in the elements A(M, N) and A(N, M) with due regard to the specified $x-y$ symmetry. Line 110 avoids the case $(M, N) = (M_0, N_0)$. Lines 80 to 100 set the range of the M and N indices in accord with the prescribed parities, ensuring that elements which should be zero are left untouched and that if PX = PY elements with N < M are not treated, since they can be filled in by symmetry. If PX ≠ PY then all the A(M, N) are scanned.

Lines 140 and 150 work out the revised eigenvalue estimate. The R value of 0.3 can be adjusted to help attain convergence in any particular case.

Lines 200 to 205 work out F(M, N), making use of the precomputed elements of the G array.

The program as presented here has no explicit statements which initialize elements to zero, except for those in line 10: a dimension declaration in BASIC automatically sets all elements to zero. When the manual command GO TO 70 is used to change the number of basis functions, that number should not be decreased, since incorrect non-zero edge elements will then remain in the A(M, N) array and will affect the subsequent calculation. The usual procedure is to *increase* Q, to check whether the calculated eigenvalue has reached its asymptotic value. The stored A(M, N) values then help to speed up convergence and the edge elements still have the zero values which were set in the initial array declaration.

```
  7 REM  ***********************************
  8 REM  TWODOSC
  9 REM  ***********************************
 10 DIM A(40,40): DIM G(50): DIM D(80)
 20 FOR M=0 TO 49: LET G(M+1)=(M+1)*(M+2): NEXT M
 29 REM  ***********************************
 30 INPUT "XY PARITY,1 EVEN,-1 ODD";PXY
 40 INPUT "PX,M0,PY,N0";PX,M0,PY,N0
 45 LET A(M0+4,N0+4)=1: LET A(N0+4,M0+4)=PXY
 47 REM  ***********************************
 48 REM  USE GOTO 50 MANUALLY
 49 REM  ***********************************
 50 INPUT "V2,BETA";V2,BE: LET B=BE*BE-V2
 60 FOR M=0 TO 79: LET D(M+1)=2*BE*(M+1): NEXT M
 69 REM  ***********************************
 70 INPUT "LAMDA";L: INPUT "DIM";Q: INPUT "E";E
 77 REM  ***********************************
 78 REM  RELAXATION SCAN
 79 REM  ***********************************
 80 FOR M=PX TO PX+2*(Q-1) STEP 2
 90 LET NM=M+2*(PXY=-1): IF PX<>PY THEN LET NM=PY
100 FOR N=NM TO PY+2*(Q-1) STEP 2
110 IF M=M0 AND N=N0 THEN GO TO 130
120 GO SUB 200: LET S=F/(D(M+N+1)-E): LET A(M+4,N+4)=S: LET
A(N+4,M+4)=S*PXY
130 NEXT N: NEXT M
137 REM  ***********************************
138 REM  ENERGY ESTIMATE
140 LET M=M0: LET N=N0: GO SUB 200: LET EP=D(M0+N0+1)-F
150 LET R=.3: LET E=R*EP+(1-R)*E: PRINT E: POKE 23692,2: GO TO 80
197 REM  ***********************************
```

```
198 REM  F SUBROUTINE
199 REM  ********************************
200 LET F=G(M+1)*A(M+6,N+4)+G(N+1)*A(M+4,N+6)+B*(A(M+2,N+4)+
A(M+4,N+2))-L*A(M+2,N+2)
205 RETURN
206 REM  ********************************
```

TWODOSC. Specimen results

10.6 Results at small λ

Following the notation used in the perturbation calculation of Killingbeck and Jones (1986) we describe the low-lying states of the perturbed oscillator by using symbols such as (m, n, e) and (m, n, o). The letters e and o mean even or odd with respect to the $x \leftrightarrow y$ interchange, while m and n are the harmonic oscillator quantum numbers associated with the product state which is the eigenfunction in the limit $\lambda \to 0$. For small λ it turns out that the choice $M_0 = m$ and $N_0 = n$ is appropriate to pick out a particular state using the algorithm of TWODOSC. At $\lambda = 0$ the choice $\beta = 1$ is appropriate, and the first few unperturbed eigenfunctions (when correctly symmetrized) consist of the exponential term multiplied by the following factors:

$$
\begin{array}{ll}
(0, 0, e) & 1 \\
(0, 1, e), (0, 1, o) & x \pm y \\
(1, 1, e) & xy \\
(0, 2, e), (0, 2, o) & (x^2 - \tfrac{1}{2}) \pm (y^2 - \tfrac{1}{2}).
\end{array}
\tag{10.9}
$$

The states $(0, 1, e)$ and $(0, 1, o)$ have mixed parity (PX $\neq$ PY) and remain degenerate as λ is increased; in group theoretical jargon, they belong to a two-dimensional irreducible representation of the symmetry group of the Schrödinger equation. The three states $(1, 1, e)$, $(0, 2, e)$ and $(0, 2, o)$ start off degenerate at $\lambda = 0$, but split up into three separate levels as λ increases. Table 10.1 shows the results given by

Table 10.1 TWODOSC results with $\lambda = 0.1$, $\beta = 1$.

State	Q = 6	7	8
(0, 0, e)	(2.0241)383	383	383
(0, 1, e)	(4.0708)643	639	639
(1, 1, e)	(6.2082)964	901	909
(0, 2, e)	(6.1592)748	868	856
(0, 2, o)	(6.0716)357	405	400

TWODOSC at $\lambda = 0.1$. The Q value was increased in steps of one by using the manual commands GO TO 50 or GO TO 70.

The results show an alternating convergence for the last three states as Q increases, but if β is increased to 1.1 the convergence becomes monotonic downwards; this feature has been found to persist through a variety of test calculations. The calculation for the five lowest states was carried out up to $\lambda = 1$ (where a β value of 2 is appropriate) and it was found necessary to go up to or beyond $Q = 15$ to obtain *explicit* asymptotic eigenvalues. This involves a lengthy calculation. However, it was observed empirically that the last three digits of the eigenvalue can be obtained accurately by Aitken extrapolation of the sequence of eigenvalue estimates obtained as Q is increased in unit steps. This cuts down the computing time required, and also suggests that for sufficiently large Q the error is proportional to $\exp(-\alpha Q)$ for some parameter α. A full Padé analysis of the sequence of eigenvalue estimates is also possible, and yields good results even when the last three or four digits have not yet explicitly converged. As examples for the eigenvalues at $\lambda = 1$ we may quote the values 2.195 918 2 for $(0, 0, e)$, 6.557 803 3 for $(0, 2, o)$ and 7.444 581 7 for $(1, 1, e)$. A calculation with $\lambda = 1$ and $V_2 = 0$ was also tried, in an attempt to find the ground-state energy for the potential $x^2 y^2$. This was estimated by Vrscay and Handy (1989) to be 1.110 ± 0.002. Using TWODOSC and WYNN we obtained the estimate 1.108 22, while applying the inner product method of Killingbeck and Jones (1986) gave 2.108 21.

ZEEMAN. Mathematical theory

10.7 Introduction

The non-relativistic Schrödinger equation for the hydrogen atom perturbed by a magnetic field which is directed along the z axis takes the form

$$-\tfrac{1}{2}\nabla^2\psi - r^{-1}\psi + \tfrac{1}{8}\gamma^2(x^2 + y^2)\psi + \tfrac{1}{2}\gamma l_z\psi = E\psi. \qquad (10.10)$$

In (10.10) atomic units are used, with the unperturbed ground-state energy being $-\frac{1}{2}$ and with the magnetic field strength γ expressed in units such that $\gamma = 1$ corresponds to a magnetic field of 2.3505×10^5 T. The z component of angular momentum, l_z, gives a simple additive energy contribution which is linear in γ, but it is the quadratic term which causes the major mathematical difficulty in calculating accurate energy levels. When γ is small (i.e. $\gamma < 0.1$) it is possible to obtain fairly good estimates of the energy levels using the program RADIAL together with some mathematical principles explained by Killingbeck (1979b, 1981c). For the spherically symmetric s states the principle of the method is easy to follow. For such states the expectation value $\langle x^2 + y^2 \rangle$, which leads to the first-order energy, is equal to $\frac{2}{3}\langle r^2 \rangle$. If the function $x^2 + y^2$ is written as

$$x^2 + y^2 = \tfrac{2}{3}r^2 - \tfrac{1}{3}(3z^2 - r^2) \tag{10.11}$$

it is clear that the second term has zero expectation value for any s-type function. However, when the first term only is used to replace $x^2 + y^2$ in the Schrödinger equation, we get a radial problem which can be solved using a variety of accurate methods. In terms of perturbation theory the dominant perturbing term is being treated to infinite order (i.e. exactly). If the second term in (10.11) is then added in as the final small perturbation it produces a zero first-order energy shift, and the second-order shift which it produces can be estimated accurately using the Hylleraas variational principle (Killingbeck 1979b). In applying that principle it is only necessary to know quantities such as $\langle r^N \rangle$ which are available from the results of the previous calculation for the radial problem. The calculation is a good example of the symbiotic relationship between algebraic manipulation and numerical computation; recently Schmidt and Nicolaides (1989) have developed the method up to the fifth order of perturbation theory for the 1s ground state.

10.8 Implicit basis methods

If the calculation outlined above is treated as a matrix diagonalization calculation, it can be seen to involve as basis functions the eigenfunctions of a *perturbed* radial Schrödinger equation. These functions can only be found numerically. Fortunately, it is only expectation values which need to be calculated. To establish this requires (as usual) a little algebra. Starting from the Schrödinger equation

$$H\psi = -\alpha \mathrm{D}^2\psi + V\psi = E\psi \tag{10.12}$$

it is possible to derive the result

$$\langle \psi | F_j[H, F_k] | \psi \rangle = \alpha \langle \psi | \operatorname{grad} F_j \cdot \operatorname{grad} F_k | \psi \rangle \tag{10.13}$$

which relates expectation values taken with respect to the eigenfunction. In (10.13) the set of functions F_j are functions of the space coordinates only and $[H, F_k]$ is a commutator. From the result (10.13) and the identity

$$F_j H F_k = F_j[H, F_k] + F_j F_k H \tag{10.14}$$

it is possible to obtain the matrix elements of H between the functions $F_j\psi$ and $F_k\psi$. If an extra perturbing potential U is added to the Hamiltonian H, the entire matrix of the perturbed Hamiltonian can be expressed in terms of expectation values taken with respect to ψ. Further, if the F_j are chosen to be simple functions of r^N type, with V a function of r only, then the hypervirial theorems which were used in HYPOSC can be used to relate the various $\langle r^N \rangle$ values, thus reducing the number of expectation values which require explicit numerical computation. This method of calculation, which we can call the implicit basis approach, has the amusing feature that the basis functions themselves need not be specified or calculated, since only expectation values are used. These can be calculated by means which involve only eigenvalues, as has been made clear at several points throughout this book. For the ground state of the Schrödinger equation (10.10) the implicit basis set $F_k = r^k(3z^2 - r^2)\psi$ would be appropriate (with $k = 0, 1, 2$, etc) in the Hylleraas principle, ψ being the eigenfunction obtained with the radial function $\frac{2}{3}r^2$ replacing $x^2 + y^2$. The appropriate decomposition which replaces (10.11) for p and d states was given by Killingbeck (1981c).

10.9 Using explicit basis functions

The implicit basis method has the obvious property that the basis functions vary with γ. While this makes them more effective, it also means that the matrix elements have to be computed afresh if γ is varied. The usual procedure is to use a basis set which has no *explicit* γ dependence, although it might contain variable parameters for which the best value will turn out to depend on γ. For the Zeeman effect problem described by the Schrödinger equation (10.10) the most popular basis functions are hydrogenic orbitals or ones closely related to them (Praddaude 1972, Brandi 1975). Edmonds (1973) and Clark and Taylor (1982) recommended the use of a basis of Sturmian functions, which gives a banded generalized matrix eigenvalue problem. Very accurate eigenvalues were obtained by Rosner *et al* (1984) using a method which starts from a set of coupled differential equations for the different angular momentum contributions to the wavefunction. The program ZEEMAN gives energies for small γ and for low-lying states which are as accurate as the ones which they obtained. It uses a basis set of functions which are equivalent to a Sturmian basis set but which are not orthonormal. The method of calculation used is a novel combination of shooting and relaxation. It is a special adaptation of the technique used in TWODOSC to a situation in which the dominant contributions to the wavefunction all belong to one value of angular momentum, with only small corrections coming from basis states with other angular momenta.

The wavefunction ψ is taken to have a form which involves the spherical polar coordinates r, θ and ϕ:

$$\psi = \exp(-\beta r) \sum W(N, L) r^N Y_L^M. \tag{10.15}$$

Y_L^M is used to denote the product of a factor $\exp(\mathrm{i}M\phi)$ and an associated Legendre polynomial $P_L^M(\theta, \phi)$. The next step is to substitute the postulated form for ψ into the Schrödinger equation (10.10) and deduce the recurrence relation obeyed by the coefficients $W(N, L)$. The algebra is more lengthy than that required in the case of TWODOSC, and requires the use of the recurrence relation obeyed by the P_L^M (Arfken 1985)

$$(2L + 1)\mu P_L^M = (L + M)P_{L-1}^M + (L + 1 - M)P_{L+1}^M. \tag{10.16}$$

Here μ denotes $\cos\theta$; the $x^2 + y^2$ term in the Schrödinger equation can be written as $r^2(1 - \mu^2)$. The terms involving the kinetic energy operator can be simplified by using the result

$$\nabla^2(r^N Y_L^M) = (N - L)(N + 1 + L) r^{N-2} Y_L^M. \tag{10.17}$$

The recurrence relation finally obtained for the $W(N, L)$ is

$$\begin{aligned} &\tfrac{1}{2}(N + 2 - L)(N + 3 + L)W(N + 2, L) \\ &\quad = [\beta(N + 2) - Z]W(N + 1, L) - [E + \beta^2/2]W(N, L) \\ &\qquad + (\gamma^2/8)[D(L)W(N - 2, L) + A(L + 2)W(N - 2, L + 2)] \\ &\qquad + B(L - 2)W(N - 2, L - 2) \end{aligned} \tag{10.18}$$

where the A, B and D coefficients are defined by the equations

$$(2L + 1)(2L - 1)A(L) = -(L + M)(L + M - 1) \tag{10.19}$$

$$(2L + 1)(2L + 3)B(L) = -(L + 1 - M)(L + 2 - M) \tag{10.20}$$

$$D(L) = -A(L) - B(L). \tag{10.21}$$

The magnetic quantum number M (i.e. the l_z eigenvalue) takes a definite value for each eigenfunction. It appears in the values of the A, B and D coefficients, but for brevity has not been used as an index in the W array elements or in the recurrence relation.

The recurrence relation (10.18) can be rewritten as a matrix eigenvalue problem, and the iterative method used in TWODOSC immediately suggests itself as being applicable here also. However, inspection of the recurrence relation shows that the approximation of setting the A and B coefficients equal to zero leaves a problem which involves only one fixed angular momentum L. That problem is just the one which would result if the potential $x^2 + y^2$ were replaced by some multiple of r^2, that is it is the effective potential problem discussed by Killingbeck (1979b, 1981c). Since that problem can be treated using RADIAL or SERAT, it is easy to check the eigenvalues arising from (10.18) when the A and B

coefficients are omitted, and so to ensure that the technique used in applying (10.18) is satisfactory. To apply (10.18) some initial L value L_0 is chosen and all $W(N, L_0)$ with $N < L_0$ are set equal to zero. $W(L_0, L_0)$ is given some large fixed value such as 1E30, since the $W(N, L)$ fall off rapidly as N and L increase. The recurrence relation (10.18) is used to calculate the $W(N, L_0)$ for $N > L_0$, up to some maximum N value NU, for a trial E value. E is varied to make $W(NU, L_0)$ zero, in what is essentially a shooting calculation with an associated root-finding module. As NU is increased the eigenvalues obtained tend to asymptotic values; these were found to agree with the energies given by the other methods for the effective radial potential. To include the other angular momenta $L_0 + 2$, $L_0 + 4$, etc in the calculation a relaxation stage is incorporated between each shooting scan. The recurrence relation (10.18) is rearranged so that the $W(N+1, L)$ term appears on the left; this allows the $W(N, L)$ for $L \neq L_0$ to be calculated, for $NU > N > L$, with all other $W(N, L)$ for $L \neq L_0$ being held at zero. In this shooting–relaxation process, the values produced by shooting in the L_0 column diffuse outwards through the (N, L) array and then produce a back-coupling which drags the shooting eigenvalue away from its original value. Eventually the whole $W(N, L)$ array settles down to a stable distribution; the effect of adding one angular momentum value at a time to the basis set can be studied by including one more column at a time in the relaxation process. It would, of course, be possible to make the entire calculation a relaxation one, by using relaxation instead of shooting in the L_0 column. This works, but test calculations indicated that it tends to converge to ground states of each symmetry type (i.e. states $1s_0$, $2p_0$, $2p_{-1}$, $3d_{-1}$, $3d_{-2}$) whereas with shooting it is also possible to obtain results for excited states (e.g. $2s_0, 3p_0, 3p_{-1}$).

ZEEMAN. Programming notes

10.10 Array requirements

The array $W(N, L)$ is two-dimensional. Strictly speaking, each element of it depends on M also, but this M dependence is taken care of by the M dependence of the coefficient arrays, A, B and D, which are computed before the shooting–relaxation process begins. The array indices are shifted upwards by four to avoid zero indices, since both N and L in the original algebra can have zero values. In a shooting calculation the wavefunction near the top end of the range can be unstable, even though the eigenvalue is accurate; this was explained in connection with FIDIF and RADIAL. As an *ad hoc* allowance for this, the NU value for L_0 has been made greater than that for the other L

values, to avoid unstable values being pumped into the relaxation process. The limiting eigenvalues will not be affected by this; an alternative procedure would be to include the last few elements of $W(N, L_0)$ in the relaxation process, with $W(NU, L_0)$ rigidly held at zero.

10.11 Subroutine structure

Two subroutines are used, one for the shooting process with $L = L_0$ and the other for the relaxation process over all the other L values. The shooting subroutine requires some kind of root-finding module to control it and to produce the eigenvalue. In the program shown here a crude Newton's method root finder is used; this was found to be adequate.

10.12 ZEEMAN. Program analysis and program

Line 10 sets up the required arrays. Fixed dimensions have been used here, with specified subarrays being picked out for use by later dimension specifications.

Line 20 requires the operator to specify the unperturbed values of the magnetic quantum number M and angular momentum L_0. M is fixed for the eigenstate concerned, whereas the perturbation mixes other L components into L_0.

Lines 30 to 70 precompute the various coefficients used in the recurrence relation for the W(N, L), using X, Y and D as array names.

Lines 100 to 130 are input lines to specify the β parameter and the magnetic field strength γ. Lines 110 and 130 convert these quantities into the appropriate quantities needed in the recurrence relation.

Lines 150 and 160 set the upper limit NU on the N values to be used in the W(N, L) array, and set the fixed initial element equal to 1E30. NUU in line 150 is the *ad hoc* augmented NU value used in the L_0 shooting process, as explained in the programming notes.

Line 170 sets the initial energy estimate EI and also the quantity IM which (in line 320) controls how many L contributions are included in the relaxation process. IM must be chosen to be an *even* integer.

Line 180 subtracts from EI the linear magnetic contribution, so that it is the quadratic Zeeman term for which the calculation is performed. In line 250 the linear term is correctly added back again when the energy value is displayed.

Lines 200 to 270 give a Newton's method control module for the shooting process, with a fixed DE value of 0.001. The I loop (lines 200 and 270) could be repeated more than once if required, although this has not been found to be essential. The Sinclair BASIC POKE in line 260 is not needed for other microcomputers.

Line 300 sets up the L_0 column for the corrected E value.

Lines 310 to 380 control the relaxation process. Line 310 sets the number of relaxation scans between each shooting scan. Making more than one relaxation scan helps to give a quicker settling down of the W(N, L) values as they diffuse outwards from the L_0 column. In line 320 the range of L values to be used (see line 330) is determined. Line 320 assumes the convention that if IM is zero the loop is *not* performed at all. An alternative would be to have an extra line of form

```
315 IF IM=0 THEN GO TO 370
```

Lines 400 to 420 are the relaxation subroutine, which use the recurrence relation to work out the W(N, L) for the L value specified in line 330.

Lines 500 to 560 are the shooting subroutine, which is used for $L = L_0$, as set in line 210. Note that the recurrence relation is used differently in the two subroutines. In the shooting subroutine the *highest* W(N, L) is found in terms of the lower ones, while in the relaxation subroutine a W(N, L) value is found in terms of the values of other W(N, L) which surround it.

To increase the value of NU or to include more L components the manual command GO TO 100 suffices. It should be noted, however, that (just as for TWODOSC) no explicit zero initialization has been included in the program. This means that correct results are only obtained if NU and IM are *not decreased* during a particular calculation, since otherwise non-zero elements around the edge of the arrays will interfere with the W(N, L) values obtained.

```
 7 REM  ***********************************
 8 REM  ZEEMAN
 9 REM  ***********************************
10 DIM W(100,20): DIM D(20): DIM X(20): DIM Y(20)
17 REM  ***********************************
18 REM  PRECOMPUTE COEFFICIENTS
19 REM  ***********************************
20 INPUT "M,L0";M,L0
30 FOR L=0 TO 16
40 LET X=(L+1-M)*(L+2-M)/((L+L+1)*(L+L+3))
50 LET Y=(L+M)*(L+M-1)/((L+L+1)*(L+L-1))
60 LET X(L+4)=-X: LET Y(L+4)=-Y: LET D(L+4)=X+Y
70 NEXT L
97 REM  ***********************************
98 REM  USE MANUAL GOTO 100
99 REM  ***********************************
```

```
100 INPUT "BETA";BE
110 LET E0=BE*BE/2
120 INPUT "GAMMA";GA
130 LET LA=GA*GA/8
150 INPUT "NU";NU: LET NUU=NU+5
160 LET W(L0+4,L0+4)=1E30
170 INPUT "IM";IM: INPUT "E";EI
180 LET E=EI-M*GA/2
197 REM  **********************************
198 REM  NEWTON SHOOTING SCAN
199 REM
200 FOR I=1 TO 1
210 LET K=1: LET L=L0
220 GO SUB 500: LET G=F
230 LET E=E+.001: GO SUB 500
240 LET E=E-.001+.001*G/(G-F)
250 PRINT E+M*GA/2
260 POKE 23692,2
270 NEXT I
297 REM  **********************************
298 REM  RELAXATION SCAN
299 REM  **********************************
300 GO SUB 500
310 FOR J=1 TO 3
320 FOR I=2 TO IM STEP 2
330 LET L=L0+I
340 FOR N=NU TO L-1 STEP -1
350 GO SUB 400: NEXT N
360 NEXT I
370 NEXT J
380 GO TO 200
397 REM  **********************************
398 REM  RELAXATION FORMULA
399 REM  **********************************
400 LET S=(E+E0)*W(N+4,L+4)-LA*(D(L+4)*W(N+2,L+4)+X(L+2)*
W(N+2,L+2)+Y(L+6)*W(N+2,L+6))+W(N+6,L+4)*(N+2-L)*(N+3+L)/2
410 LET W(N+5,L+4)=S/(BE*(N+2)-1)
420 RETURN
497 REM  **********************************
498 REM  SHOOTING SUBROUTINE
499 REM  **********************************
500 FOR N=L-1 TO NUU
510 LET S=-(E+E0)*W(N+4,L+4)+LA*(D(L+4)*
W(N+2,L+4)+X(L+2)*W(N+2,L+2)+Y(L+6)*
W(N+2,L+6))+(BE*(N+2)-1)*W(N+5,L+4)
520 LET D=(N+2-L)*(N+3+L)/2
530 LET W(N+6,L+4)=S/D
540 NEXT N
```

```
550 LET F=W(NU+6,L+4)
560 RETURN
561 REM  ***********************************
```

ZEEMAN. Speciman results

10.13 Shooting–relaxation mode

Results for several eigenstates were given by Killingbeck (1987f) using a program of shooting–relaxation type closely similar to ZEEMAN, and ZEEMAN reproduces those results. As an example, we show the results for the ground state to illustrate how the results vary as NU and IM are increased. L(max) in table 10.2 is the maximum L value used in the basis set.

Table 10.2 ZEEMAN eigenvalues for the 1s state.

L(max)	γ 0.1	0.2	0.3
	(−0.49)	(−0.49)	(−0.47)
0	752165	031737	892850
2	752648	038148	918558
4	752648	038156	918654
6	752648	038156	918655
(β, NU)	(1, 10)	(1, 15)	(3/2, 25)

The results are very accurate, but care has to be taken in the choice of β in order to obtain convergence of the iterative process. The reader may have noted the unorthodox way in which the relaxation is performed; $W(N+1, L)$ rather than $W(N, L)$ is evaluated, so that the divisor $\beta(N+2)-1$ increases with N, if β is kept sufficiently large. The 'obvious' choice of the element to evaluate, $W(N, L)$, is not so good here, since it involves the small N-independent divisor $E+\beta^2/2$.

Bibliography

Aasen J O and Romberg W 1965 *B.I.T.* **5** 221
Aitken A C 1932 *Proc. Edinburgh Math. Soc.* **3** 56
Albasiny E L and Hoskins W A 1969 *Comput. J.* **12** 151
Allen G D, Chui C K, Madych W R, Narcowich F J and Smith P W 1974 *Bull. Aust. Math. Soc.* **11** 63
Andrew A L 1986 *B.I.T.* **26** 251
Andrew A L and Paine J W 1985 *Numer. Math.* **47** 289
Arfken G 1985 *Mathematical Methods for Physicists* (New York: Academic Press) p669
Baker G A 1965 *Adv. Theor. Phys.* **1** 1
Banerjee K 1978 *Proc. R. Soc.* **A364** 265
Barakat R and Rosner R 1981 *Phys. Lett. A* **83** 149
Barth W, Martin R S and Wilkinson J H 1967 *Numer. Math.* **9** 386
Beech G 1980 *Successful Software for Small Computers* (Wilmslow, England: Sigma Technical Press)
Ben-Israel A 1966 *Math. Comput.* **20** 439
Bickley W G 1968 *Comput. J.* **11** 206
Birkhoff G, De Boor C, Swartz B and Wendroff B 1966 *SIAM J. Numer. Anal.* **3** 188
Blankenbecler R, De Grand T and Sugar R L 1980 *Phys. Rev.* D **21** 1055
Boys S F 1950 *Proc. R. Soc.* **A201** 125
Brandi H S 1975 *Phys. Rev.* A **11** 1835
Buenda E and Guardiola R 1985 *J. Comput. Phys.* **60** 561
Carter S and Handy N C 1987 *Comput. Phys. Commun.* **44** 1
Chang J, Moiseyev N and Wyatt R E 1986 *J. Chem. Phys.* **84** 4997
Channabasappa M N 1979 *B.I.T.* **19** 134
Chaudhuri R N 1985 *Phys. Rev.* D **31** 2687
Clark W C and Taylor K T 1982 *J. Phys. B: At. Mol. Phys.* **15** 1175
Cocke J and Kennedy K 1977 *Commun. ACM* **20** 85
Cohen A M 1980 *Int. J. Comput. Math.* **8** 137
Collar A R 1948 *Q. J. Mech.* **1** 145
Common A K 1968 *J. Math. Phys.* **9** 32
Cooley J W 1961 *Math. Comput.* **15** 363

Coquereaux R, Grossman A and Lautrup B E 1990 *IMA J. Numer. Anal.* **10** 119
Crandall S H 1951 *Proc. R. Soc.* **A207** 416
Devries P L and George T F 1980 *Mol. Phys.* **39** 701
Duck W 1964 *Z.A.M.M.* **44** 401
Edmonds A R 1973 *J. Phys. B: At. Mol. Phys.* **6** 1603
Evans D J 1973 *Comput. J.* **18** 70
Evans G A, Hyslop J and Morgan A P G 1983 *Int. J. Comput. Math.* **12** 251
Fack V, De Meyer H and Vanden Berghe G 1986 *J. Phys. A: Math. Gen.* **19** L709
Faddeeva V N 1959 *Computational Methods of Linear Algebra* (New York: Dover)
Feenberg E 1948 *Phys. Rev.* **74** 206
Fernandez F M, Meson A M and Castro E A 1985 *J. Phys. A: Math. Gen.* **18** 1389
Feshbach H 1948 *Phys. Rev.* **74** 1548
Fox L 1967 *Comput. J.* **10** 87
Frantz D D, Herschbach D R and Morgan J D 1989 *Phys. Rev.* A **40** 1175
Fyfe D J 1969 *Comput. J.* **12** 188
Ghezzi C and Jazayeri M 1982 *Programming Language Concepts* (New York: Wiley)
Gibbons A and Rytter W 1988 *Efficient Parallel Algorithms* (Cambridge: Cambridge University Press)
Ginsberg C A 1982 *Phys. Rev. Lett.* **48** 839
Goult R J, Hoskins R F, Milner J A and Pratt M J 1974 *Computational Methods in Linear Algebra* (London: Stanley Thornes)
Grant I P and Burke V M 1967 *J. Comput. Phys.* **2** 277
Greville T N E 1970 *Math. Comput.* **24** 179
Gries D 1981 *The Science of Programming* (New York: Springer)
Grogono P and Nelson S H 1982 *Program Solving and Computer Programming* (Reading, MA: Addison-Wesley)
Grosjean A and Jolicard G 1987 *J. Phys. B: At. Mol. Phys.* **20** 4635
Guttmann A J 1977 *Programming and Algorithms* (London: Heinemann)
Hajj F Y 1982 *J. Phys. B: At. Mol. Phys.* **15** 683
Hamer C J and Barber M N 1981 *J. Phys. A: Math. Gen.* **14** 2009
Handy C R and Williams R M 1987 *J. Phys. A: Math. Gen.* **20** 2315
Harel D 1987 *Algorithmics* (Wokingham: Addison-Wesley)
Hautot A 1982 *J. Comput. Phys.* **47** 477
Hautot A and Nicolas M 1983 *J. Phys. A: Math. Gen.* **16** 2953
Håvie T 1966 *B.I.T.* **6** 24
—— 1979 *B.I.T.* **19** 204
Hikami S and Brezin E 1979 *J. Phys. A: Math. Gen.* **12** 759
Hoare C A R 1969 *Commun. ACM* **12** 576
—— 1986 *BYTE* August p115
—— 1987 *Computer* September p85
Hyslop J 1972 *Comput. J.* **15** 140
Jahn H A 1948 *Q. J. Mech.* **1** 131
Jamieson M J 1983 *J. Phys. B: At. Mol. Phys.* **16** L391

Johnson B R 1977 *J. Chem. Phys.* **67** 4068
Jolicard G and Grosjean A 1985 *Phys. Rev.* A **32** 2051
Jolicard G and Perrin M Y 1989 *J. Chem. Phys.* **91** 7780
Kahaner D K 1972 *Math. Comput.* **26** 689
Kantaris N and Howden P F 1983 *The Universal Equation Solver* (Wilmslow, England: Sigma Technical Press)
Kemble E C 1958 *The Fundamental Principles of Quantum Mechanics* (New York: Dover)
Khalifa A K A and Eilbeck J C 1982 *IMA J. Numer. Anal.* **2** 111
Killingbeck J P 1975 *Techniques of Applied Quantum Mechanics* (available from author)
—— 1977 *J. Phys. A: Math. Gen.* **10** L99
—— 1979a *Comput. Phys. Commun.* **18** 211
—— 1979b *J. Phys. B: At. Mol. Phys.* **12** 25
—— 1980 *J. Phys. A: Math. Gen.* **13** L35
—— 1981a *The Creative Use of Calculators* (Harmondsworth: Penguin)
—— 1981b *Phys. Lett.* A **84** 95
—— 1981c *J. Phys. B: At. Mol. Phys.* **14** L461
—— 1985a *Microcomputer Quantum Mechanics* 2nd edn (Bristol: Adam Hilger)
—— 1985b *Rep. Prog. Phys.* **48** 53
—— 1985c *J. Phys. A: Math. Gen.* **18** 245
—— 1985d *J. Phys. A: Math. Gen.* **18** L1025
—— 1986a *Phys. Lett.* A **115** 301
—— 1986b *J. Phys. A: Math. Gen* **19** 2903
—— 1987a *Eur. J. Phys.* **8** 196
—— 1987b *J. Phys. A: Math. Gen.* **20** 601
—— 1987c *J. Phys. A: Math. Gen.* **20** 1411
—— 1987d *J. Phys. A: Math. Gen.* **20** 1285
—— 1987e *Eur. J. Phys.* **8** 268
—— 1987f *J. Phys. B: At. Mol. Phys.* **20** 5387
—— 1988a *Phys. Lett.* A **132** 223
—— 1988b *J. Phys. A: Math. Gen.* **21** 3399
—— 1988c *J. Phys. A: Math. Gen.* **21** 111
—— 1988d *J. Phys. A: Math. Gen.* **21** 1917
Killingbeck J P and Galicia S 1980 *J. Phys. A: Math. Gen.* **13** 3419
Killingbeck J P and Jones M N 1986 *J. Phys. A: Math. Gen.* **19** 705
Killingbeck J P, Jones M N and Thompson M J 1985 *J. Phys. A: Math. Gen.* **18** 793
Knuth D E 1974 *ACM Comput. Surv.* **6** 261
Kohn W 1949 *J. Chem. Phys.* **17** 670
Krogh F T 1970 *Math. Comput.* **24** 185
Kuo C 1971 *Computer Methods for Ship Surface Design* (London: Longman)
Levin D 1973 *Int. J. Comput. Math.* **3** 371
Longman I M 1971 *Int. J. Comput. Math.* **3** 53
Löwdin P O 1951 *J. Phys. A: Math. Gen.* **22** 4089
Lyness J N 1972 *B.I.T.* **12** 194
Lyness J N and Ninham B W 1967 *Math. Comput.* **21** 162
Makarewicz J 1989 *J. Phys. A: Math. Gen.* **22** 4089

Marsden M J 1974 *Bull. Am. Math. Soc.* **80** 903
Mathur S N and Harmony M D 1977 *J. Chem. Phys.* **66** 3131
Mills H D 1975 *Commun. ACM* **18** 43
Myers G Y 1975 *Reliable Software through Composite Design* (New York: Van Nostrand Reinhold)
Nash J C 1979 *Compact Numerical Methods for Computers* (Bristol: Adam Hilger)
—— 1990 *Compact Numerical Methods for Computers* 2nd edn (Bristol: Adam Hilger)
Nesbet R K 1965 *J. Chem. Phys.* **43** 311
Neville E H 1934 *J. Indian Math. Soc.* **20** 87
Nisbet R M 1972 *J. Comput. Phys.* **10** 614
Penrose R 1955 *Proc. Cambridge Philos. Soc.* **51** 406
Peters G and Wilkinson J H 1969 *Comput. J.* **12** 398
—— 1970 *Comput. J.* **13** 309
—— 1975 *Commun. ACM* **18** 20
Porter M B and Reiss E L 1986 *SIAM J. Numer. Anal.* **23** 1034
Powell M J D 1981 *Approximation Theory and Methods* (Cambridge: Cambridge University Press)
Praddaude H C 1972 *Phys. Rev.* A **6** 1321
Pythian J E and Williams R 1986 *Int. J. Numer. Methods Eng.* **23** 305
Ralston A and Rabinowitz P 1978 *A First Course in Numerical Analysis* (Tokyo: McGraw-Hill Kogakusha)
Rao S S 1989 *The Finite Element Method in Engineering* (Oxford: Pergamon)
Rellich F 1969 *Perturbation Theory of Eigenvalue Problems* (New York: Gordon and Breach)
Rice J R 1983 *Numerical Methods, Software and Analysis* (Tokyo: McGraw-Hill Kogakusha)
Richardson L F and Gaunt J A 1927 *Trans. R. Soc.* **A226** 299
Romberg W 1955 *K. Nor. Vidensk. Selsk. Dorh.* **28** no 7
Rosner W, Wunner G, Herold H and Ruder H 1984 *J. Phys. B: At. Mol. Phys.* **17** 29
Rutishauser H 1963 *Numer. Math.* **5** 48
—— 1966 *Numer. Math.* **9** 1
Sale A 1975 *Aust. Comput. J.* **7** 116
Schmidt H M and Nicolaides C A 1989 *J. Phys. B: At. Mol. Opt. Phys.* **22** 1751
Schultz G 1933 *Z.A.M.M.* **13** 57
Schwartz C 1969 *J. Comput. Phys.* **4** 19
Secrest D, Cashion K and Hirschfelder J P 1962 *J. Chem. Phys.* **37** 839
Shammas N C 1988 *BYTE* September p295
Shanks D 1955 *J. Math. and Phys.* **34** 1
Shishov V A 1961 *Zh. Vych. Mat.* **1** 169
Shore B W 1973a *J. Chem. Phys.* **58** 3855
—— 1973b *J. Phys. B: At. Mol. Phys.* **6** 1923
Silverman J N 1983 *J. Phys. A: Math. Gen.* **16** 3471
Smith W E and Lyness J N 1969 *Math. Comput.* **23** 231
Sparks M 1970 *Q. J. Appl. Math.* **28** 103
Squire W 1975 *Int. J. Comput. Math.* **5** 81

Talman J D 1980 *J. Comput. Phys.* **37** 19
Tanenbaum A S 1976 *ACM Comput. Surv.* **8** 155
Tewarson R P and Zhang Y 1986 *Int. J. Numer. Meth. Eng.* **23** 707
Turing A 1948 *Q. J. Mech.* **1** 287
Usmani R A 1987 *B.I.T.* **27** 615
Vanden Broeck J M and Schwartz L W 1979 *SIAM J. Math. Anal.* **10** 658
Vorobyev Y V 1965 *Method of Moments in Applied Mathematics* (New York: Gordon and Breach)
Vrscay E R and Handy C R 1989 *J. Phys. A: Math. Gen.* **22** 823
Wilkinson J H 1965 *The Algebraic Eigenvalue Problem* (London: Oxford University Press)
Williams A R and Weaire D 1976 *J. Phys. C: Solid State Phys.* **9** L47
Wirth N 1973 *Systematic Programming* (Englewood Cliffs, NJ: Prentice-Hall)
Wynn P 1956 *M.T.A.C.* **10** 91
—— 1966 *Numer. Math.* **8** 264
Yourdon E 1975 *Techniques of Program Structure and Design* (Englewood Cliffs, NJ: Prentice Hall)
Zave P 1984 *Commun. ACM* **27** 104
Znojil M 1976 *J. Phys. A: Math. Gen.* **9** 1
—— 1981 *Phys. Rev.* D **24** 903
—— 1986 *Phys. Lett.* A **116** 207

Index